THE CONSTRUCTION
OF
PHYSICAL AND EMOTIONAL HEALTH

The Construction
of
Physical And Emotional Health

Sergio López Ramos

Library of Congress Control Number: 2012906596
ISBN: Hardcover   978-1-4633-2682-1
                         Softcover   978-1-4633-2681-4
                         Ebook   978-1-4633-2683-8

This book was printed in the United States of America.

To order additional copies of this book, contact:
Palibrio
1663 Liberty Drive, Suite 200
Bloomington, IN 47403
Tel: 877.407.5847
Fax: +1.812.355.1576
orders@palibrio.com
401178

Table of Contents

To: Erik Sosa Cobos
Who has seen the sunrise of the King Sun
in the pyramid of sun
and made him a bow

Acknowledgements

At the Universidad Nacional Autónoma de México (National Autonomous University of Mexico), School of Advanced Studies Iztacala for being my alma mater, my fellow seminar,El cuerpo humano y sus vericuetos. (The human body and its intricacies), to meet more than 17 years working as a team, Margarita Rivera Mendoza,Norma Delia Duran Amavizca, Irma Herrera Obregon, Araceli Silverio Cortes, Rosa Sosa Consuelo López, Gerardo Aguilera Abel Chaparro Veronica Diaz, ZendoTeotihuacán as a space for inner growth and Jagüey Institute where I learned a lot about cancer and chronic degenerative diseases, as patients and have accompanied me all these years. A special mention and my gratitude to Pável López Ozuna y Adriana Vega Vega, for his entire good job in the completion of this English translation. I hope to give some hope to those seeking a light on the way and be a different look to the unborn.

Teotihuacán
Spring of 2012

Introduction

Thirty years ago I joined the Universidad Nacional Autónoma de México (National Autonomous University of Mexico) as a teacher of psychology. As a result of spreading the education project in the context of the government of Luis Echeverría, were created the *Escuelas Nacionales de Estudios Profesionales Aragón, Acatlán, Cuautitlán, Iztacala y Zaragoza* (National School of Professional Studies Aragon, Acatlan, Cuautitlan, Iztacala and Zaragoza). I studied in Iztacala and then I settled down there. Years later those schools became Faculties and they remain the same up to now. This comes about because in 1978, was the behaviourist hegemony school and did not allow other thoughts or psychological theories on campus, while the population increased demands on their conditions, economic and cultural poverty seemed to have no limit, today there are millions who still lack access to health, culture, housing, recreation, etc. We can not say that they have access to their body, their processes and messages that build on these days the threat of death in the disputes of drug markets nationally and internationally for groups of drug traffickers and political power, the anxiety and the culture of fear have taken over the body and no matter the social status the transformation of family and personal life is real hard to hide, that resulted in new patients, with bodies full of allergies, psychosomatic problems, chronic degenerative diseases, cancer, sleep disorders, gynecological problems, sexual identity issues, depression, anorexia, bulimia, among many others, the social reality makes us more and more artificial. We move away from our nature and live mental ideas we build or were built by the mass media, the representation of life, we are filled with worldly desires that never end, therefore, we cultivate the suffering and so even though we have no physical condition, we have an emotional condition because we have a desire, we want to have more or care more about what we have, the case is that we enter into a dynamic of pain for desire and then we build a condition.

It was very much reality for a behavioral theory, which presented its first crisis in the year 1981. We wanted to change the curriculum but nothing was done. The import of psychological models was in crisis, its epistemological and philosophical explanations did not correspond with the emergence of our reality, fewer strategies to solve the psychological problems of Mexicans who cannot learn or receive therapy while they are hungry or unemployed.

This means that the human body is going to build alternatives to protect life, the presence of self-regulatory mechanism in us a chance to think about other alternatives, think about that possibility when there was no solution to the orthodoxy of a psychology that just wants condition or verbalize. While the patient cannot find a way out of his body process that does not understand, because it crosses the guilt and sin and feels punished, we find a body is entering a new response for survival, therefore, must have its history, was not the result of divine punishment, nor was evolution, less a machine, we found it a body-building process that relates to the history of the society in which we live, the culture of physical neglect, with the depreciation of not deserving anything, with the loss of affection and esteem, it weighs in memory that we are no good for nothing and that we are an inferior race, finding that with ancestors is a severe shock, keeping out of the emotional state that is said that we are a sad country reason why we eat too much spicy (chile).

What psychology could give us another look to solve or prevent physical, psychological and psychosomatic problems of our patients, none gave us a viable, all aimed at specialization and fragmentation of the human being, so it had to be trained academically in other fields of knowledge, we seek national history, literature, alternative medicine, acupuncture, anthropology, philosophy, epistemology, work with the body moving cellular and emotional memory, zazen meditation, among other things and we started to apply them to patients, there were good results, hence derive the need to theorize about the etiology of psychosomatic problems, which are increasing, along with cancer and chronic degenerative diseases.

Performing the work of deep reflection of body process demanded to create its history, how it is instituted in a society and become "normal" in the eyes of generations, to unravel the process made us get out of the canon of psychological science that only search stimuli or causes and use other disciplines such as history, oral history, anthropology, acupuncture as a theoretical model, philosophy and epistemology to build another interpretation of the process in front of us, a body with multiple choices of notice. Our first explanation was to differentiate between categories of body and corporal for contending that the subject makes the process of building a space, geography, culture, religion, food systems, a family history and last but not least their body also builds solutions to conditions of danger to the life of the body. To discover and document this reality, took us several years, we made the history of psychology in Mexico, the history of the body, documentation and illustration of the corporal and psychosomatic process, we have come to volume VII, we create workplaces, try to illustrate this fact in many ways, in newspapers, in stories, novels, life stories, many senior thesis, master's and doctorate, as a result of that background the texts were born and make up this book.

I made an arrangement out of sequence in production, I made three sections in the first one named *The background*, I try to answer that question, why do the subject of the body, secondly, *The problem* shows the vision of work and its epistemology in a kind of interpreting the corporal (physical) that manages to look beyond frontiers to positivism, the third, I named; *The explanation and solution*, I try to go to the daily part of the body that named psychosomatic can be found in any area of the body and to culminate a work with diabetic women.

Thus, the paperworks are a sample of the process that I discovered in practice and theory, as they were scattered in different books and hard to find, some sold out. I brought them together in a way that allow a kind of synthesis, to understand our theoretical proposal and I would not like to detach them now, I think it's better for the reader to reveal his own truth, make his own judgments and expressions, and say, "this is me" or "how do you know", if that happens I think I will have fulfilled my mission.

Zendo Teotihuacán
Winter 2011

I

The background

1

The Historical and the physical: ideas and reflections

The science is our concern, psychology has become part of a discussion about what it means to be a discipline that is built with the processes of other disciplines, its set of truths becomes a story of horror and disappointment about the truth of the human condition. Because when it seemed that the truth was staying in the labs and its light illuminated the reason as absolute certainty in human being, that moment was seen as the possibility to build and complete the final version on the knowledge of man. Fortunately always appeared a doubt or a critical light of reason and was as faded as a fantasy that is shared collectively. There was only the record of a science that began the climb to regain the truth in man kind; explore his body, the unconscious, language or behavior only represents the duality of extremes on what involves an exploration of fragmentation and unity of divisible; knowledge could only be hold on the idea of analogy and fragmentation as principle, the elements of nature were called for predictable, this means self-affirmation.

The history of science is a good instrument to open some interpretive horizons to make some proposals about knowledge in a particular context, especially the geography of the responsibility to consider the beginning of time and space, in a proposal looking ahead, with a significance that involves thinking about a reality. In other words, culture as the principle that articulates meanings and courses of lifestyles. The construction of historical knowledge as a possibility is just a place where reading immediate places changes its intensity and shape.

The depth of what you want to build goes through this process of knowledge of those places where one prostrates, and it involves an interpretation or look at our own identity. The principle is an element that must specify an action to break the fragmentary reading of the human and place it in the eyes of the historic construction.

History, the historical-social, two moments of the same circumstances. Historic becomes an opportunity to reflect that crosses boundaries of all knowledge and human action. Therefore, speaking of the historical, this time allows me to turn around what I do and what others think about the work of psychological knowledge. I'll take the historical-social resource to be an interpretive tool that breaks with the dogmatism and atomization, its conjugation with the culture, anthropology, semiotics and the universe of inter subjective relations are possibilities that break with the absolute truth and categorical statements. Lead to a reality that is diluted with a subjective world that seeks security in what is a false security that can not be built with the policy of the unions and gregarious sense of self-protection, resulting in the defense of spaces and institutional theories to the defense, regardless of social and cultural processes that are the source of its configuration, the closeness to reality is diluted and removes all traces of history to human action, which involves avoiding man can build, survive to get by, the homo-academic seeks security.

Cartesian reason taught him to drive away ghosts principle of intuition. Standardization as a possibility for all people only leads to unified language and stereotyped forms of repetition, which validate the principles of universal laws. In that logic makes no sense to doubt the truth and become universal truths. It hurts to say they will never be. But that only shows that the idea of dominating nature becomes a slogan sick and predatory. No matter the space where they live or who live there. If that happens with the immediate geographic area, what can you expect about the human body or, in other words, the duality that is constructed as an option within the truth seen as a vision of what happens inside and outside of human beings. Fragmentation is not only a principle, which spreads the world of words and their meanings; no, is a concrete action that is embodied in lifestyles and who makes the epistemology of science.

Its realization is beyond the philosophical discourse, the fears of living prostrate in the body memory and thus are organic processes that do not fit into the logic of fragmentation. The human body is diluted as amorphous idea that makes no sense in the proposal to think that things can be articulated outside of logic, looking for relationships where none exist. The depth of what I say is not a sign to disqualify. My attitude towards different thinking that can penetrate the narrow world of psychology. Of course, everyone finds what you want and makes sure the plurality of ideas and freedom of research, only with that reading can justify countless atrocities committed in the name of science, I think the chances of going deeper into reflection is just the beginning of construction that gives us the pattern for interpretation and readings different from our present interests. The body as a unit that does not allow the fragmentation and divisions for the sake of functionality will not search for meaning and importance of living in a body with deep historical and cultural roots. I can not close the possibility of reading, accept that human being is a microcosm that holds a deep analogy with the earth is a correlation that can change the reading of what is called psychological and therefore we can build new relationships inside and outside the body, that is to say, seeing it as a unit that is built into the culture, geography and weather conditions of territorial areas. Knowing that the human body and the historical become inseparable condition can alter us the importance of making the interpretations we make to patients and epistemological reflections of present to planning the uncertain future.

I think the value of a discussion gives us the opportunity to understand how the body is constructed and obtained some autonomy or became the tyrant disputing the organs, thereby causing life forms to fill of a principle on the invalidity of a time that is structured as a chance to do readings convictions, which leads me to wonder why psychology has become a technocratic tool to make predictions self- affirmative.

All this provides the possibility of interpreting the process of building the body in a conjugation of search of an articulation with a unit called the human body. I turned to history, acupuncture and the practice of zazen. The joint points offer the possibility of constructing an interpretation going beyond the appearance, an artificial reality becomes an obstacle to interpret the facts and ways of life that we face daily. No doubt the epistemological problem is more complex in the eyes of positivism, because it reached its interpretive limits and does not respond to a human body that is blurred to fragmentation that sought in the recesses of the cell and microbiology, as which did not allow scientists to consider the unity of the process of building the human body and its multiple processes, which do not give a single interpretation to the ways of living. That allowed seeing that affirmative borders in psychology were not conclusive and this opens new hopes of doing psychology.

The practice with patients allows to rethink the concept of the body and what builds the subject; internal processes according to their conscience and the ways to build according to their culture and ways to make the existence in coordination with the lifestyles and family history; the result, a story that is confronted with a new physical reality, which is blurring the proposal to seek only the emotional and allowed me to find new relationships in the body, constructed by the subject, see a historical-cultural-nutritional-sentimental-hatred-hard feelings-complaining about life process, combined with a dose of stress, all of which the body becomes a contraction and permanent deformation, with a process manager obsessions, addictions, a body memory that transforms gradually and results in lifestyles and ways of being in the individual. The process however, did not end there, the combination of the above involves a more complex relationship and we are still working to understand how an individual builds symptoms and signs that show the breakdown of a principle of harmonic balance between organs, leading to a deterioration of physical and emotional health. What does this mean? There is a link between organs and that one can be more dominant than another and this may be the beginning of a construction that overlooks the ways that subjects live. We can say that someone is more liver and thus, among other relationships, is a very scary or sentimental.

The complexity of this approach allows joint lifestyles, ways to express emotions and explanations that are built in accordance with cultures and geography to which it belongs. This is for us a new opportunity to understand the body as a microcosm unit that can articulate other relationships and interpretations of what is considered the psychological. I do not think that this reflection allows to understand the significance of psychology as a unique style of approaching to the body, the fragmentation gestates a serious problem about the epistemology of the emotions and psychology: how to solve the problem of duality of Cartesian thought. For us that is not a problem, we do not accept such a division because it is artificial, built to the poverty of physical reality: a thought that is deeply fearful in the exploration of reality can not be constructive in proposing the future. Consider any theory of history of science and we can see that see the past as continuity or discontinuity does not mean that it constitutes a watershed in social life, no, it's just the beginning that allowed us to isolate and fragment an idea about the construction of human being. Today is laughable to hear that "you are supposed to do the search for unity, of integral" when the unit is in reality but not in our brain. Part of the problem are the eyes-culture that we see the object and there is the division for purposes of convenience or is a line of autism that does not socialize new views about the object. The last exploration we do not know today is the conclusion of the genome project, which is a living example of the atomization of Cartesian knowledge and is the loss of the human condition at its best.

However, some researchers in this field agree that they do not know what happens in that long chain inhabited by chromosomal genes, but the opening of the genetic information has a close relationship with the emotions and feelings and not just proteins, such as geneticists say. The concept that we handle, we cannot ignore the principle of internal relationship of emotions and organs. It is a dance of energy balance that may be changed by an emotion or a feeling that creates forms of communication without letting open up new humanized possibilities, but harmful to the life processes. The worship of destructive processes is part of this process where life is not respected. It's worth noting that competition becomes a social practice and individual culminating in a degenerative process of health, both physical and emotional; enough to see the bodies of people to understand that emotional and organic processes go hand in hand, not only positivist logic, but physicalism and orthodox Marxism in some cases are sentencing and absolutist, passing by psychoanalytic statements that subscribe to sexuality or that want to reduce all human behavior to an instinct. One cannot be so sharp with regard to life of human beings. But the truth is that competition becomes destructive of life expectancy and that does not feed the spirit, that is, without becoming more human and that's a crucial point in our day. The emptiness is not resolved with the therapies that we know in the middle, you have to go a little beyond the simple statements about what it means to the interior or spirituality, culture of resignation becomes a possibility of existence and gives the body a rhythm, attitude and above all a way of life, with a bodily life that is not built with a nice lifestyle, it is only an intention which can not let the individual grow. The crisis of values would not only be explained as a redoubt of social decay but interior of human beings.

This is present in the body memory of the human body; the progress of physics in medicine have found that this idea focused on brain memory becomes a museum piece to the holographic proposal of the body and its memory. The combination of the universal in the particular proposal embodies the microcosm that is governed according to the principles of rotation and translation of the planet, including the analogy with the Earth, on the composition of water and earth, the bones and days of the year, the relationship with the day and night to communicate with others, and so on. Certainly, for us is a field that has allowed to make major changes in our vision and professional practice, the results have been satisfactory in relation to the other and ourselves. We believe that the psychological cannot be a combination of refined duality now seeks its origin in the particles that make up the fragmented view of the past three hundred years. Statistics can be a good reason to start thinking about what the dualistic view has done in psychology today.

2

Why the story of the human body in Mexico

The aspect of most discussion in history is the boundary of what is understood as the true story and the story reported by the political group in power. Everything is built on individuals, the social and geographical identity, a culture of everyday life. Clifford Geertz[1] spoke of building the web where materialize the ideas of existence is a country if we go further, give meaning to the collective imagination and feed the possibility of political and cultural hegemony, opening the eyes of citizens to the official history means baring the society of its history and feeding the belief that everything in society is natural and there is a way of life and death which is not disputed, the completion and extension of this concept in our grandparents and parents are irrefutable evidence: history objectified in a corpse, a dead body[2]

---

[1] Clifford, Geertz, *La interpretación de las culturas*. Gedisa, Barcelona, 1995,p.387.

[2] I mean a loss of memory in society today where we cannot find continuity to the dead process. The pain of loss can not go to the analysis of how they die, you can not find the historical strands of historical process that is present in a loved one because they look beyond our reality or our time: What has 1870 to do with me, I live in 2012?
Surely for the laziest of mortals nothing. But for dealing with the process of seeking the intricacies of the web, whether a relationship exists.

Even dead, the human body is the document that can be studied by anthropologists, forensic scientists, archaeologists, historians, among other disciplines, it all depends on what you will find how to build a concept of the body in the Mexican society. Its story is pigmented by religious and scientific policies[3], of deliberate actions against social and racial groups that are seeking new possibilities to disqualify the excluded, their bodies will not be equal, they would not have the same forms and ways of eating, even a dead body would be an extension of other bodies, rituals may range from burial to cremation. The presence of the body will be gone in living bodies of this, both in memory and in the actions of a lifestyle that will revive the immediate past of a body that does not die or disappear, diluted, absorbed, integrates, turns to concrete in the children or grandchildren of a family and that is when we see that the body is perfected or debilitate in a historical and social process that always flows according to the rules and conventions that may hinder future search[4].

Where to look? In the newspaper libraries, because they register the immediate feeling of a problem, space for gossip, of the ephemeral[5] but with the consistency to be a thought that is indicative of how it will give the process of interpreting the human body and which is the remaining. The documentary first-hand source allows new possibilities of reading with the present and therein lays our search on the proposal to see the past of

---

3   Tate Lanning, John, *El real protomedicato. La reglamentación de la profesión médica el imperio español*, Facultad de Medicina. Instituto de Investigaciones Jurídicas, UNAM, México, 1997, p. 569.

4   Rico Bobio, A. *Teoría corporal del derecho.* Miguel Ángel Porrúa. Universidad Autónoma de Chihuahua, 2000, p.232.
    The author says that the body is "the totality of a being articulated in this case the human, in all aspects visible and invisible, perceptible or inferable only by indirect means," p. 80. I infer that it is not feasible in our case, we can go back and rebuild there.

5   McGowan, Gerald, *Prensa y poder*, Colmex, México, 1978, p.376.
    Makes a study on the subject of the press, which states that the notion expressed in the papers are, valid opinions in a moment. Tracing these ideas shows the thinking of an era. For our case, we went to the newspapers because they are a source that gives opportunity for continuity of the ideas expressed and inform us about the culmination and the institution of the subjective criteria of truth.

the human body as a critical and proactive ability to see in the present work systems in the health field[6]. Another reason for the history of the human body in Mexico focuses on the construction of options to interpreter in the present on the health field of different sectors of population and to know who decided what was good for the body. What is the right thing in health? What is the best way to die and who should and must rule on the truth and the true in the body? Many answers are in the past and has remained in the present, we can even say that the body and health become a source of books and research that feed a line and confirm an idea about health and their immediate and future implications. But the body seems to be diluted and does not exist, all proposals are aimed at diseases, hygiene, application of medicines, surgery, public health, etc., this means that an individual appears to be amorphous for receiving and prescribing all kinds of solutions, regardless his gender or age, much less his body[7]. Research on health or disease focus on the institutional

---

[6] Health systems today face one of its most acute crisis in terms of service forms and efficiency of treatments. What has led to some rethinking of traditional forms of healing, the resurgence of alternative medicine responds to the high rate of iatrogenic and irrational use of drugs that produces a high percentage of deaths in hospitals.
See: Contreras, Francisco, *Salud en el siglo; XXI ¿sobrevivirán los médicos?* Book, Crafer, México, 1998, p.451. The averages are 27% for Mexico.

[7] In this respect the literature is abundant: Flores, A. Francisco, *Historia de la Medicina en México. Desde la época de los indios hasta la presente.* Oficina tipográfica de la Secretaria de Fomento, México, 1886, tomo I y II. Con prólogo de Porfirio Parra. That is a positivist history of medicine in Mexico. But it is rich in data and treatments, especially on therapeutic: Chávez, Ignacio, *México en la cultura médica.* Instituto Nacional de la Salud Pública, Fondo de Cultura Económica, Colección biblioteca de la Salud, México, 1987, p.147. Where the concern of working with the body does not appear in the story it tells, there are the institutions or the concept of health that is driven to make sense of their stories: Cañedo, Luis et al, *La salud de los mexicanos y la medicina en México.* Editorial del Colegio Nacional, México, 1977, p.482. "La población del país es el sujeto del problema de salud" (at p.8). Certainly is not the body seems that the intention is to give a subject who has no surname least one body. No doubt I'm doing a reading that finds the body where there is no object, but this view loses sight of the possibilities of social and individual construction of the subject so his body does not appear

aspect and that makes its object to be centered in the search of multiple correlations about services and ways to make health care even get to the aspects of service evaluation when there is no decline in death rates or epidemiological problems. The body is diluted in service policy and ingestion of drugs, seems not to exist[8].

This change of writing history about the human body has not been known in the national history of medicine, however, based on our historical and anthropological readings we can reflect on the uses of a concept of body. In the history of medical power instituted[9] we observe the continuing trend towards the body fragmentation that gives users a reading that takes them away from the appropriation of their feelings. The emotional and physical part as an integral element of existence, the distance of the body becomes a search that does not contact with what is experienced in the body, fragmentation is a fact of everyday life that attempts to explain the rational thinking, divide, not only has the purpose of study also involves a process of dismantling the reality of individuals, both public and private, are created consensus, hegemony and a look that allows monocausal readings, the appropriation by the individual institutes in the body the conditionality that becomes a criterion of truth in physiological functioning, this means that the body is not a biological

---

in the treatments, everything is standardized or required to regulate and the individual is only a figure in a chart or a report.

[8] A text that touches the aspect of evaluation and organization is Donabedian Avedis, *Los espacios de la salud; Aspectos fundamentales de la organización de la atención médica,* Secretaria de la salud, Instituto Nacional de la Salud Pública y Fondo de Cultura Económica, México, 1988,p.771. Even the concern centers on the customer's appearance, provide a quality service, but do not appear. Authors such as Evans G. Robert., et al *¿Por qué alguna gente está sana y otra no? Los determinantes de la salud en las poblaciones.* Organización Panamericana de la Salud, 1994, p.411. At most they get is to consider the cultural and what is its impact on health.

[9] López Ramos S. *Prensa, cuerpo y salud, en el siglo XIX mexicano (1840-1900)* Miguel Ángel Porrúa CEAPAC, México, 2000, p.353. Doctors settled in the political power from the years of 1847 and since then drove their fragmented reading of the human body, even the negation of human subjectivity is related to the positivist philosophy, which was established in the medical cradle of thought of XIX century.

reality, is a historical result, is a complex process[10] that, demands the use of the history as a tool that allows us to give new meaning and dimension to the processes. The human body is not exempt from this condition.

The multiple subjective and organic interrelations built by individuals as well as the ways in which they appropriate them, give a different dimension to the ways of living, that is, building new networks of cooperation or survival inside the organs of the subject[11]. The combination with ignorance, malnutrition and overwork provide us with bodies living with panic and fear, circulating in their cells.

That, sometimes not allowing being in the here and now, that is, know where I stand as a character of society and how I should participate.

The bodies are consumed, are transformed by generations of undernourishment and the history of geography: it becomes underdeveloped, a few years of life. That was the XIX century in Mexico. These are some of the reasons that led me to search the immediate step, how is established in the body the policy of a government, to unravel the dark side of the steps and reflections on a social group to another, resize a document written between 1846-1900 with the instruments of present and for other eyes that can give the required reading according to their questions.

These documents, which I analyze, are part of that, to find how to build socially and historically the body and, beyond that, the source, the

---

[10] I can even say that depredation of the planet has its bodily effects, because the body is part of a broader process than culture, that is, is articulated as climate change and soil disturbance. This means that what happens in the immediate environment has consequences in the medium and long term in individuals, so the one to one relationship cannot be correlated immediately. This is detailed in "Notas para la historia del aire en D.F.", en novelo Victoria y López Ramos Sergio, *Etnográfica de la vida cotidiana*. Miguel Ángel, Porrúa, México, p.82-92.

[11] Capra, Fritjof, *La trama de la vida. Una nueva perspectiva de los sistemas vivos*. Anagrama, Barcelona, 1999, p.359. The use of force as a principle of survival, have passed away now. Today people are more competitive and cooperative which enable them to survive longer and this attitude continues in the legacy, and how it works the body as a memory creates new learning networks and cooperation. This is a trend of individuals today.

document and its interrelation to cultural and scientific aspects of a time allow new interpretations of a body in a lifestyle that disrupts the threshold of present. The need of century XXI to see the past as an instrument that could be the possibility of understanding the present and thereby to blur the established ignorance in the myths and statements being touted for the new generations about their bodies using inheritance genetics as a subterfuge of historic evasion. We can be certain that a review of social and political processes in the body, in Mexican society, from official medicine are fraught with doubt on the statement of the quality of life, especially when we know the mortality rates in our days and cure systems are located in a causal physical reality which does not respond to a concept of integral or unit with a macrocosm. Epistemological problems and intervention facing official medicine today seem to have no other way to accept the status of this concept only led to the disappointment of not finding the truth of life or eternal youth[12].

With the articles I found, in the XIX century, one can say: this style of living and dying has its history, it is not our natural bodily existence, the correlation with social process is increasingly undeniable even by the positivists, is no longer possible to say that a disease is caused by an isolated body. Today the explanatory range of health includes emotional, sentimental aspects and direct correlation with the physiological processes[13], is a unit that has always existed but that the positivists of the

---

[12] It should be noted that the last year's reflections on health considerations have diversified and soft medicines (massage, acupuncture, herbal medicine, naturopathy, chiropractic etc.) have been gaining ground. Especially because they were considered that the body is vital space to recover health and when treatments are individualized, treatments and diagnoses are more suited to the subject and circumstance that although it has become fertile ground for some doctors who continue with the idea of fragmentation and want to poach on the field as their own. Some texts are: Munné, Antoni, *La evidencia del cuerpo; como llegar al equilibrio cuerpo-mente-espíritu*, Paidós, España, p. 346; Rivas Vilchis, J.F. *Acupuntura y plantas medicinales*. Editorial Herbal, México, 1999 p.212. Hage, Mike, *El libro de dolor de espalda*, Paidós, España, 2001; Schjelderup, Vilhelm, *La nueva medicina*, Ed. Miraguano, Barcelona, 1984,p.254.

[13] Martín Paul, *Enfermar o curar por la mente. El cerebro y el sistema inmunitario*. Temas de debate, España, 1997, p.501.

XIX and XX centuries refused to explore for the sake of their exploration methods could not go beyond the tangible or we could say that their instruments were not giving them answers about the internal microcosm of the human body, but still, did not respect and understand its dimensions, it is not just enough to see inside an object of study, you need to see the correlations and how are inscribed in a geographical culture or planet. That means being able to conceive the dimension of a body and the life forms involved and the effects it has on him and vice versa, this reasoning could not do the positivists of the past centuries, they were against the cell, DNA, genetic inheritance, but in the end something was needed to understand this dimension of the body in a broader context, what this means, simply that it bare of emotions, feelings, spirit and had before it a piece of cells that do not move by themselves, are related to that emotional state to create new internal networks in the body, therefore, the universal laws failed, there are unique individuals and can not be standardized or the symptoms, nor the physiological processes, the construction of an individual is given in the society in which he lived, however, things are not that simple, the subject may not realize this process, "the truth" established by scientists is an important factor in this process should happen too many things in his life and especially must have an explanation that allows release of the official proposal, but unfortunately it is not commonly found as well, gets the idea: the way things are, that's life. He resigns himself to live and die like all his ancestors.

For these reasons it is valuable to go back in time to unravel the institution of a concept and its implications in the present, for our case is the body that is constructed and operating or becoming the subject of discussions and science policies having no other concern that the market dominance and political control, build an image of the poor and criminals from your body is a social segregation that prevents a sector of society to move up the social ladder, Do they feared the Spanish and Creoles? Of losing the economic benefits that the natives claim their land if they learned to read and write, this idea is best understood when it prevents eating meat and amaranth so they can not grow, produce malnutrition has a profound political meaning into believing that they are inferior bodies have their emotional effect on self-esteem and minimizes the expectations to climb the social ladder. Knowing that policies for the identification of prisoners were intended for the Indians and the poor only polarized social differences, this policy was not far from racist proposals of the English and French[14], because a modern society encloses the beggars and panhandlers, their bodies were not (are) worthy show of modernity and the institution of a body to the poor and one for the rich is a distinction of social evolution, social Darwinism of the XIX and XX centuries found justifications in practices seeking a guilty of course this logic is far from a historical process of human beings, they saw them as something natural and normal.

Our sources reveal a process conjugated with physicians, scientists, journalists who promoted the ideas of scientists in order to tell the truth, today, the process is the same just saying from a recognized authority by a group that is the truth about a certain topic, even though they say it was false afterwards. In the XIX century was experienced this process of ideological struggle between liberals and conservatives, a hot topic in the year 1846 in the matter of phrenology, the discussion focuses on knowing where the spirituality of the human soul is, the argument of materialists,- positivists should be understood- was that "the brain digests impressions of the senses and transforms them into ideas[15]." One of the members is focused on where there is the process of thinking, soul and brain act stated: "Two are their fundamental principles:

1) Soul, mind or understanding, operating through the brain.
2) The soul has different powers that are manifested through the corresponding brain organs[16]" Phrenology became the proposal that gives opportunity to make room for the existence of spirit. "What's New in Phrenology, unlike physiology is distributed to the brain and various parts that were particular organs of various

---

[14] Comas, Juan, *Las primeras instrucciones para la investigación antropológica en México*. 1862, p. UNAM Cuadernos del Instituto de Historia, México, 1962, p.43. This was a manual to identify the Indians in the New World, prepared by the French.
[15] "Argumentos de los materialistas tomados de la Frenología contra la espiritualidad del alma racional y su solución." En *El católico*, núm. 8, pp.15, 170-72, Sábado 18 de Abril de 1846.
[16] Ibid pp.170-172.

faculties of the spirit[17]" There were the arguments to typify the body and soul of human beings, only that the main division was given with the brain, this focus gives differential treatment to the rest of the body and is defined as the organ allowing one thing to communicate with another. While the rest of the body goes to the level of support or provider of a brain that allows the soul and the brain communicate via the language[18].

Everyone got up with their truth on the body, health and ways of dying, even the request for death penalty is a discussion for solving the process of rebels[19] of the soul, had a more practical and builder sense of body from individuals, it was said in the *Spectator of Mexico*[20] whose mission is only to explain the causes of various ills and commodities, and gradually go through the human condition, obtaining the perfect harmony with religion, morality and reason ". That teaching was aimed at building a basic principle in the existence of human knowledge, the harmony of the triad had a deep religious sense, of course, to moralize the body had (have) a sense that defines the boundary and the examples were irrefutable samples for these conservatives say, the matter of these doctors and their role with the individual's body and behavior:

"What matters is materialistic or atheistic, who knows its duties demanded by his profession? He is not responsible for making men but cure them. Will not be easy to answer this miserable objection. The medical profession is a kind of priesthood, whose performance must have the heart, the intention, eyes, hands and pure human being, compassionate, and discreet. These virtues are incompatible with the doctrines of materialism. Man who senses to govern, senses to satisfy, is man always ready to respect the decency? The one judging that pleasure and pain are the

---

[17] "Argumentos de los materialistas tomados de la Frenología contra la espiritualidad del alma racional y su solución." En *El católico*, Sábado 25 de Abril de 1846.num. 9,pp.17,194-6.

[18] "Argumentos de los materialistas tomados de la Frenología contra la espiritualidad del alma racional y su solución." En *El católico*, núm. 11, pp.21, 242-44, Sábado 9 de Mayo de 1849.

[19] "Pena de muerte" *El monitor Republicano*, Octubre 9 de 1849.

[20] "Frenología", en *El espectador de México*, núm., 9 de junio de 1851.

only difference between vice and virtue, Will man be scrupulous or will abuse on confidence received? Man who sees man as a vicious animal, as being sensitive only? Will man know to feel pity on our miseries and perform his duties with that sweet and indulgent humanity, which often replace the impotence of art for the consolations of charity? Man who has no conscience is he faithful in keeping the secrets of families, which often is depository[21]?

The distance between the atheist and believer seems to dispute that seems ideological and not scientific, the truth was that the debate marked a distance about dealing with doctors and body concept looks like it has to repair the object, even the same supporters of phrenology say it is beyond the man-machine relationship. A new player joins in the discussion of phrenology. Francisco de Castro and Barcelo, said his phrenology was a spiritualist, but the uses were very clear: "We have also examined in this City and in front of the honest, intelligent and high social position, eight heads of criminals and results have corresponded exactly with the crimes that have justified this is proof that truth can practice a phrenologist[22]. "Without fear of ridicule, Castro and Barceló makes his approach in terms of correlation studies to do with the body and the actions of individuals were shown irrefutable science that phrenology was positive use to society, the protection of the honest people, intelligents and high social position demanded a science that was accurate in the diagnosis and could leave in jail to criminals or thieves by hunger. There was one thing against him, the deformity resulting from the lack of nutrients, education without the need to express to anyone and to top it off their color, their lack of religiosity, it must be said that these skulls were measured of the poor, the author gave his argument to make clear to the men who criticized the fundamental principles of phrenology spiritualism:

1 Faculties of the soul are innate.
2 The spirit works through the medium of material organs.
3 The brain is the organ of the soul.

---

21 ibid.
22 Francisco de Castro y Barceló. "Frenología, en *El monitor Republicano*, México, p. 17 de julio e 1851."

4   The soul rules the body; the material is subjected to the spirit.
5   The brain is not a simple organ.
6   The size of a brain organ, other things being equal is the positive measure of its mental power.
7   All faculty of the soul, when this predominantly active, has its special language and nature.
8   The size and shape of the brain, is distinguished by the size and external shape of the skull or head[23].

One thing was the statement of principles and other the uses to measure the subject's body to find a relationship with a behavior, racism was not explicitly constructed if not arguments that gave the appearance of science, not racism selection of social groups. The body of the inhabitants of this land the rising Mexican Republic had three blood groups: *a, o y b*, product of a mixture and a political marriage between the indigenous, Spanish and blacks, make a selection on the social ladder of evolution had a specific social function and could not escape from it. The blissful spiritual phrenology only confirmed his conviction based on body typify and cranial, Francisco de Castro and Barcelo made a dissertation on a series of articles to convince those who deplored his proposals in a mixture of arguments about the functions of brain and the role of phrenology and the benefits this brought to society in his time. Its authors are Cubi, Gall drivers of phrenology, he comes to a conclusion that links the functions of the body by the governing body, the brain, "We believe, therefore, have established the plurality of the organs of the brain, where resulting two bodies can not occupy the same place simultaneously they should be located at different sites: from here comes the location. "

The Spiritualists have done an objection lately: the unity of ego that our intelligence cannot reconcile with the plurality of the organs.

As if we were accustomed to see bent the order of our understanding of all kinds of things discussed here but, What to say to the undoubted duplication in each direction and unity of perception, because we do not see double, between the diversity of the senses and the homogeneity

---

23   Francisco de Castro y Barceló. "Frenología", en *El monitor Republicano*, México, p. 20 de julio de 1851."

of sensations? The soul escapes from all our inquiries and disconcerts all our combinations, or rather that in this way is shown that there is something about our vain pride and our weak reason[24]. These approaches of the author gives us an idea of his concern about the corporal and the importance of the time the idea of interpreter all through the brain as the governing organ that gives the orders and let the other forms are submitted and are even governed by this man who gives the pass to the soul and spirit. Body concept, which is not far from the fragmentation and division of the organs. Perhaps we need to keep in mind that this reading is to make a classification of physical appearance and thereby allows some men to measure their language or set aside the racial heritage to speak of a pure race. The political background of this proposal will not be different to the intentions of the groups in power, too much criticism. We had to make new proposals and led to look at other ways to make the body of "others" could be read with eyes full of racism, class or even claims of honest people and intelligent. Thus poor or thieves are fools, the label becomes in a process that can lead to social condemnation and escape from it is less possible.

The newborn Mexican society, in 1821, wanting to be cosmopolitan had to deny three hundred years of colony, as well as its dark or backward state, depending on the parameters of the incipient positive science and its concept of quality of life. The concept of body proposed in the mid XIX century only will follow the positivist pragmatism that contained an institutional structure and education with a group of scientists and physicians[25]. For that reason it was not just a relationship one by one in

---

[24]   Francisco de Castro y Barceló. "Frenología, Artículo sexto" en *El monitor Republicano*, México, p. 23 de julio de 1851."

[25]   In these years, doctors did many descriptions of deformed bodies. Its importance can be placed on the need to ensure this continuity of a concept that is derived from a logic that explores the relationship of the process. These "human monsters" I dare to say, were a sign of a food process making their toll on output growth or ingestion of any drug product. Today, the correlation with nutrition is not debatable to see children born headless. Rodríguez, Juan María, "Teratología, descripción de un monstruo humano, Diplogenésico, monocéfalo, autoritario, Onfalosito, no viable," en *Gaceta de México*, G15, núm. 10, pp.145-155, mayo de 1869; Rodríguez, Juan María, "Teratología,

everyday life to see the reciprocal body, it also needed the ideological discourse and in case you had to demarcate who was who in terms of class and culture. It stemmed from the discussion on the identity of the Mexicans, is a speech that the Spanish and their children, the Creoles developed with the intention of achieving some benefits to describe a new story and a report for the wishes of the political-economic power of Creoles converted to liberals[26]. This Social group that built the dilemma of identity to the question of not knowing whether they were Mexican or Spanish from the Iberian Peninsula. The dilemma of being or not being like slogan is concealed and diluted in the action of the War of Independence that gives a geography, a social body as possible to the need of a nation. Perhaps worth saying that the natives did not have problems with their identity, were and are Mayan, Lacandon, Yaqui, mixes, totzil, Zapotecs, etc. Children of Spanish, the Creole did, in order to access to the public and political positions of New Spain. The denial by the leaders of the Iberian Peninsula to the right to bear political power led to the independence struggle[27].

After these action began the struggle for power within the emerging Mexican Republic. In short, the problem of the corporal crosses the dispute over the political and economic power by a group in power, the Creoles. The discussion on the uses and purposes of phrenology is a proof of this, besides having the morbid curiosity of knowing who and what these measures were to locate from dangerous or potentially intelligent or murderer. Although the year of 1870 *The Republican Monitor*, on February 17, published a long text on the human face and its relationship

---

descripción de un monstruo humano, Diplogenésico, monocéfalo, autoritario, Onfalosito, no viable," en *Gaceta de México*, G15, núm. 11, pp.161-169, 1 de Junio 1869.

[26] On this subject there is considerable literature. Two texts are illustrative on this point: Brading, David, *Los orígenes sobre el nacionalismo mexicano.* México, Era; Reyes Heroles, Jesús, *El liberalismo mexicano*, Fondo de Cultura Económica, México, Tomo I, 1988.

[27] Rodríguez, O. E. *El proceso de la independencia en México.* Instituto Mora, México, 19992, 72. It is an excellent text that sets out the intentions of the independence and the costs of the armed movement by Miguel Hidalgo to the head.

with emotions and art, suggesting the influence of character expressions, opened a new field to make sense to the condition of the face in the various actions of everyday life, however, continued meetings with deformed bodies, it seemed that the identity is not built in a Mexico that has not had full face of Nation. Benito Juarez and Porfirio Diaz were locked in the dispute over the elections.

Mexico with "x" is the construction of a discourse that takes the image of an eagle devouring a snake on a cactus, the liberals arguing for a national project, for a republic of letters, for public health, for deaths reported in the newspapers daily. Stigmas which relapsed on people of dark cinnamon color were building a custom of disqualification is only one principle which bears a stamp of exclusivity to impose on the body of the other various desires and interests, without considering that the other has a different construction on what it means to live with his body. Plotinus Rodakanaty in 1874 with his publication of *El Craneoscopio* (The Cranescope) invites the return of phrenology and its application as a science that studies human nature and even proposes to install a professorship of phrenology and its application as a science that studies the nature human and even proposes to install a professorship of phrenology in Mexico City and its proposal on the body ages so:

"The ages of man, according to Gall, are five; to wit, boyhood, adolescence, youth, manhood and old age. This age variation arises from the mutation of a quality in another, leaving some time and years, and acquiring a temperament quite different. The first age is called childhood or expertise; the quality is warm and moist which lasts from birth to age 14. The second is called adolescence whose quality is hot and dry, lasting from 14 to 25 years. The third is called youth, which is very warm at first, though lasting from age 25 to 40. The fourth age is called manly and consistent quality of which is rather cold and dry, lasting from age 40 to 55. The fifth age is called senescence or old age, the quality is too cold and dry: it lasts from 55 years to the end of life. These five ages can be reduced to four being: puberty, youth, old age and decrepitude[28]. It seems that death at age 32 in the Valley of Mexico is not taken into consideration by the author, his promotion about patriotism will

---

[28] "Filosofía clásica" en *El craneoscopo*. Núm. 1 Tomo I, 29 de Abril 1874.

be a celebration of phrenology as a possibility to make Mexicans, like any animal, can inhabit a place or residence that is located in the upper brain. The expectation of exploring the brain like a space where inhabit everything man do, opens old discussions about the desire to find things resolved or cultivate them somewhere, it seems like this idea flourishes because work is not an option in the Spanish culture nor the indigenous. Plotonio intellectual curiosity does not lead to lasting proposals and their texts are part of a resurgence of the philosophy of spirituality.

We can see that the body of Mexicans is a space that is configured with the culture and ideas of others, but what is best for the citizens? In an emerging society that is projected on a new illusion, sometimes oblivious to the poor individuals who refuse to be the representatives of the "grand style" and renounce the pretended purity of the ideals of being superior, the simple life must be transformed; dress to have new attitudes and ideas about what is imagined for the body and the idea of person. The body model is an instant where individuals can be seen in a mirror. The industry in the Valley of Mexico was to change the relations of living space: there was pollution and rarefied air, epidemics decreased the body of workers and laborers, the deformations for routine work and occupational diseases will be the first impacts in a body that invaded the textile industry[29].

However, it appeared that the hegemony of a discourse is not imposed by the principle of a law or to see the rich models; the human body is the immediate realization of a culture, not its synthesis; there are other mechanisms, other processes that individual develops and makes use of the election and that is where they change, transform, enrich, new processes is brewing, new networks of survival and cooperation in the individual both organic and emotional. The worst case is the statism of resignation that involves the death of the organism[30].

And beyond the speech, geography, soil erosion, the diversion of rivers and draining of the lakes will be keeping a close relationship

---

[29] López, Ramos, S. *Historia del aire y otros olores en la ciudad de México. (1840-1900),* prólogo de Victoria Novelo, Miguel Ángel Porrúa, México, 2002.

[30] I start from the idea of F. Capra on networking to a hazardous condition of the body.

with the body forced to give new answers, to organize to give a new partnership in its interior and in its shape. There will be no conscience, only to be unpublished disorders and conditions that will challenge the rationalist thought. Searching the brain is also expressed in trying to make history epilepticus. Alvarado Miguel wrote in *La Gaceta Médica*, on December 1, 1883, his "Brief Notes to form the history of an epilepticus state" because his manifestations were recurrent and frequent, but started with data from Europe and culminates with the Mexicans keeping track of the types of demonstrations. It is an approach to the body without leaving the search scheme in the brain, which means that there is an exaggerated representation about intelligence and what human beings could develop by way of learning. Of course there is no discussion on the status of women and men, was ruled out participation of women in these matters on the brain weight and intelligence issues, certainly that is a sexist society prominently, as would say César Lombroso: "Incredibly, it is nevertheless a fact that there are no real data on the moral and physical female construction. There have been extensive studies on the Bushmen and aboriginal races of Australia, but scientifically speaking half of the human race, the fair sex admired, adored, despised and misunderstood, little is known of what is known if its country should be the Star of the Dog (Canis Major) or the planet Mars, instead of our earth[31], "If the body of men was studied for its brain, the woman as well but were made wrong statements about their weight and size, was considered mentally inferior but in surgery were superior to men, the same surgery Lombroso reports where women reported insensitivity, of course it was not lack of sensitivity in women, rather it was the doctors who operated them[32].

A body in the making, with an educational proposal that wants to be secular with Porfirio Diaz in power by the year 1889, we historically requires a reading that allows us to understand why present is an extension of the past in memory of individuals today exceed the intent

---

[31] Lombroso, César, "Estudios fisiológicos. Inestabilidad física de la mujer", en *Guía Práctica de Derecho*, 26 de octubre de 1884.

[32] López, Sánchez, O., *Mentirosas, enfermas y temperamentales. La concepción medica del cuerpo femenino durante la segunda mitad del siglo XIX en México*, Plaza y Valdés,- CEAPAC, México, 1998, 165p.

of a policy that only deals with a group and justifies what it does for the good of humanity and any atrocity occurs in the name of progress that deforms the body, subject to classification, selection, the actions of a religion castrating and denies the body in its expressions and submits it to the repression, suppression of emotions. Feelings combine the social norm of appearance and drills. It suffices that the newspaper *El Siglo XIX* *published on September 13 and 18*, 1889 two articles on "indigenous race" and showed that the degeneration of the race was due to several factors. I quote: "Early marriage almost prepubescent youth, promiscuity of the sexes in which they live and incest and consanguineous marriages are their results, are clearly a major cause of degeneration of the race, their smallness of stature, rickety complexion, the dull performance of their brain, their predisposition to certain vices and diseases, all of which decreases its activity, its longevity and reproductive output. So in the same Indian family are powerful elements of degeneration.

As for the individual, it happens the same, as well as hereditary diseases, faces the vice of alcoholism, and the bad food, almost exclusively vegetable. The Indian liquor that are ingested in the worst quality, the most toxic, being the cheapest, the most imperfectly distilled and, the most adulterated. It is known that children of alcoholics suffer the sad inheritance of epilepsy and other diseases. Regarding to food, the Indians gain real and a half daily as average wage, that has to be entirely insufficient for nutrition they rarely try meat, the corn and peppers are the main food, and we know that vegetables are not as nutritious as animal substances, highly nitrogenous. Malnourished, necessarily present less resistance to disease, and do not heal due to lack of doctors, pharmacies and resources, giving into the hands of charlatans, witches and herbalists. "

Of course, this reading is the appearance and there is no minimum critical attitude on the possible reasons to gestate the circumstance of the native born, their bodies taken to extremes of exploitation, you can say it's genocide against ethnic groups, who always carried perpetrate the slogan of taking away their land, allowing loopholes to blame them of something that violates the law. In this context, in which you must read the entitled articles *"Identification of convicted"* published by *El municipio Libre* April 21 to August 2, 1892. What may be the intention of such an explicit identification, to whom the proposal was no doubt, was for the poor and hominids, because based on the premise that the names they wore had the legacy of the stigma of evil and that all poors were potential offenders, the identification was so sophisticated that had to do a record of each part of the body to begin to classify individuals and to find correlations among offenders with a similar physical structure from head to toe through their name and family name. The body that is identified is none other than the Indians and workers accompanied with the picture. This body is enclosed emotionally and physically marked with the sign of punishment, social classes exist without undermining the policy of a democracy and the principle of respect for others, this body malnourished, diminished in the health and without expectation of compete in the social scale has to develop a system of survival inside that will give new forms of resistance to punishment. Porfirio Diaz probably was not thinking about building Lecumberri Palace to celebrate the hundred years of independence only, but to those identified had to keep them in the punishment cell which would impair their bodies and their enthusiasm for life. The proposal came from Paris. Ignacio Fernández Ortigosa was prepared expressly for such purposes; even his memory was dedicated to the General Porfirio Diaz. Complications were not only about the nation but also included the identification of security feelings of society where they lived.

There is a historical process that becomes the space where you are given the confluence of new strategies for viewing the future. Liberals and scientifics tested proposals that have their logical continuity in the public health system that leaves new questions to citizens. They were confronted with the experience of social and personal changes that are covered with the two-faced policy and belief to learn from foreigners. The extrapolation of scientific models becomes true imposing rules and regulations on what to do to track down criminals, to see individuals who are becoming frenchified. Not just enough to know that he has frenchified on clothing and political thought, and reaches beyond nutrition and lifestyle changes. A body nourished with products that belong to other territory will have its implications in the process of gestating new reactions and diseases that become more complex in an environment where epidemics become factors of change, of course in public health policy.

There is a historical process that becomes the space where you are given the confluence of new strategies for future viewing. Liberals and tested scientific proposals have logical continuity in the public health system that leaves new questions to citizens. Confronted them with the experience of social and personal changes that are bathed in the two-faced policy and belief to learn from foreigners. The extrapolation of scientific models become true that imposed rules and regulations on what to do to track down criminals, to see individuals who are becoming Francophiles. Not just enough to know that it has Frenchified on clothing and political thought, and reaches beyond nutrition and lifestyle changes. A body nourished with products that belong to other territory will have its implications in the process of gestating new reactions and diseases that become more complex in an environment where epidemics become factors of change, of course in public health policy.

We may add that Indians had no cow's milk in their memory cell, for example; hence the digestive process presents problems with lactose[33]. Therefore, in the Mexican past, we found heterogeneity of discourses and concepts that confront and defend against the new changes on health, fashion. But the question remains the same: What does it mark the body of mexicans? The epidemiological study and the relationship with the body seem not to touch eachother. The disease is an autonomous entity that deserves to be attacked from the conception of the foreign agent that invades the human body[34]. Hence, the causal reasoning that does not search the more complex processes, the emotional and organic relationship as a single entity, the fragmentation, according to positivist thought, do the dissection, a search that looks at the instance of what it should do according to the model that looks in coincidence. The effort of XIX century scientists to decipher the effects of bacteria in the body shows a constant, the use of positivist science as the only element to continue the corporal fragmentation and justify the solution and cure. The classifications

---

[33] Research on the relationship of food and quality of life have been extended to studies on blood type and pH, and consumption of certain foods that can make the blood more alkaline tissue and therefore harmful effects to health.

[34] López, Ramos, S., *Prensa, cuerpo y salud en el siglo XIX mexicano (1800-1900)*, Miguel Ángel Porrúa- CEAPAC, México, 2000. 353p. This theme is broad in chapters three and four of the cited work.

of disease allowed atomization and fragmentation of the body[35]. We understand this proposal when we know that crime rates and disease are the scourges of European cities, in the Mexican press is common to find literature of Germany, France, England and Sweden, among others, because they lead to immediate demand. And it is understandable that the search topics are extended to various fields. Texts could be read as: "Anthropology. Influence of the professions against crime." Published in *Guía práctica del derecho* (Handbook of law) in the year 1892, to assert that these reflections on crime get their influence from Paris, continuing the whiff of discrimination but now well-argued. Secundino Sosa said about epileptics:

"Trying to shape, as far as possible, medium profiled of various doctrines and most accepted that have been issued about the matter in question I realize and concrete to three and there are:

1st   The doctrine of *strict liability* in epileptics.
2nd   The doctrine of *absolute irresponsibility*
3rd   The doctrine of *diminished responsibility*.

Remained the first of these options for those days when whoever was supposed to reason was responsible, forgetting who said such acts are products of human feelings, passions, reason, conscience and will and that is exclusive absurd to take as pitch of moral the intelligence. Kept monstrous error to those times, adjacent to the other ill-fated era where epileptics and the insane were considered as being possessed out of the domain and outside the law of human sympathy[36]. "

The truth was to defend the idea of not to marry or could be reproduced for the good of society. If science had succeeded in giving place to individuals based on their appearance or their condition was and is the

---

[35]   Available at: Sergio López Ramos, Fuentes Hemerográficas para una historia del cuerpo en México, México, CEAPAC Ediciones, 2005, 353 p.

[36]   Sosa Secundino, E., "Medicina Legal. La responsabilidad en los epilépticos" en *Guía práctica de derecho*. Octubre 31 de 1892. Treaty presented at the National Academy of Medicine to qualify for the open competition for the H. Corporation. To fill a vacancy of legal medicine as an option.

criterion: the body and its presentation, in the eyes of society that wanted to be good and pretty good, Needless to say some crazy and insane would be locked in the hospitals and to prison of Belem would go those dissatisfied with the regime of Porfirio Diaz[37].

---

[37] I quote the text illustration below to express the feelings of the opponents of Diaz. "La Psicología" en *El siglo diez*, núm. 15093, 9 agosto 1888.p.1.

"Our estimable *colleague El monitor Republicano* after reproducing a clip of our newspaper entitled *Antonia* says the following:

"No, buddy, is that psychology does not play, is that terrible laws weighing on the printer press, make it almost silent. Issues to consider there is a lot of vital interest that despite the times, are initiated by the media. But writing today is so delicate that we must go very cautiously, with very great care because we already know the way to Belem."

"Sorry colleague we do not accept across the latitude these findings, that the reading could believe burden on the press a violent coercion that does not allow the slightest manifestation".

"And this is very exaggerated, since it has done more than put a stop within the law, the overflow of some newspapers that, forgetting the respect we owe to society, rushed to the field failed to slander and libel."

"The same *monitor* as a supporter of freedom of thought would never have entertained in their columns the production of weekly pamphlets that if they were not condemned by the society would be the disgrace of our time."

"We have said, and we support freedom of press, but it has no other limits than those of morality, privacy and social order, but we want these powers so the press enjoys true, not those productions that exploits the greed that hurts the editorial and public decency."

"We also remind our fellow writers often are reduced to prison, not at the request of the Government but by the complaint of some individuals who have believed offended by a publication".

"But if you look down the opposition press, can be estimated to have complete freedom to discuss the actions of the authorities, and even our fundamental laws so rudely attacked the clerical party."

"Just go to Catholic leaves persuaded that no coercion on the writers, because never in the periods in which the religious war against the Reformation caused, I raise both political hatred, was written with the virulence and rancor, with those produced today the bodies of the clergy."

"We who have the conviction that these arguments are perfectly useless, and only serve to galvanize a dead match to public awareness, several times we have encouraged the liberal newspaper to resolve issues by science and

The exercise of power by the Porfirian dictatorship includes the dimension of punishing a body that has broken the law, social norms, besides not having a type, anthropometric speaking, to be consistent with the affable presence that is not ugly or neglected and even the dress becomes a criteria to make the selection. It is history the anegdota of the calzonudos (henpecked) in the Zocalo, when modernity demanded giving work to whoever brought pants to work because the body had to be covered and the "calzonudos" were an unpleasant spectacle[38].

The dimension of the many interactions with the study of the body, not only forces us to go back to see how this logic is instituted in speeches and in everyday life, no, there are other dimensions, such as education, medical practices, the idea of making a medicine without pain, create analgesic society which can not question anything, everything tries to be a perfect circle where there are no cracks for doubt: Whoever hesitates will face the scorn, the label of charlatan and other insults. Is not that ignorance has different ways to be enlightened and live in government institutions? It's when you should look deep in speeches and history, when the claims are not correlated and ways of dying and living, morality and ethics requires us to pursue the need to know the multiple whys in a society that is combined with ancient and modern and gives us various signs of death, the quality of life is superior in its manifestations.

So in this work the reader will not find answers symmetrical, no. You can see the face of history you have in your body, school, food, customs and converted into truism, the falsehoods of the speeches do not fit into a single interpretation is required to fetch the interrelationship of a discourse

---

devote themselves only to look after the implementation only of the reform laws, trying at once to give new impetus to this, in order to put an end to violations of the principles so painfully won, and redress the wrongs that are resisting the nation for non-compliance of the Reformation."

"To encourage our colleagues to establish closer together not only the bonds of fellowship, if not the ties that should exist between the supporters of the same political creed, just look for the good of the country, so seriously threatened by the common enemy. WRITING."

[38] González Navarro, Moisés, "El Porfiriato vida social" en Cosío Villegas, Daniel (coord) Historia moderna de México III. Hermes, México, 1985, p. 396.

that has diversified and came to life itself, so the dislocation, fragmentation appears to be a natural phenomenon, this artificial disruption of the body does its work in individuals and is rarely manage to have an overall look. The intellectual effort has run into "the truth" instituted by the "experts" that clean of the criticism or doubt, and those claims become critical thoughts mowers or call into question the physical reality and the living of their counterparts. For these reasons, our work focuses on showing the veil of a body establishing the institutions and discourses of the XIX century and comes to us in the XXI with its mask of modernity and not only that, but the new resources to sell to the pores of the body, the market industry exceeded the intentions of control, it becomes thinner the process of social classes; the distinction with perfumes or clothing, and even the new body aesthetics, from extractions to liposuction, to food supplements with anabolic. Democracy or free markets are to be imposed by the ways of dealing with a body full of new eating habits deformations, from the use of hormones to fatten farm animals to so-called transgenic and irrational use of medicines, consumer societies are not only sanctuaries for bodies to die in hospitals with a ritual that degrades the one who stays alive, the dead man's impact was so violent that we can say that there is no escape. The peculiar death to each one can be quiet and full of sweetness in food; diabetes and endocrine disruption are proof of that. A body that is destroying itself not only shows the decay of a consumer society also shows that life expectancy has not been the best and that the XIX century tends its tentacles towards the cultural memory of a language that is spoken sometimes and sometimes is just postural, closely related to an ergonomic principle which doesnt see health but the market.

The above thoughts are a sign of a value of historical work and the need to see the space where we gave birth: the Mexican Republic. It turns out that this Republic has a story and the concept of body we seek from the years 1846 to 1900. Our approach is to subscribe to a period rich in struggles and contradictions of three political groups: conservatives, liberals and scientists, the last two are opened to combine and give us the inheritance of a reading of Health that continues our days.

The texts that I reviewed at that time are an amalgam of proposals and show the concern for a human being who seems not to be well identified, you can read the doubt and hesitation about what is represented as a human being, the significance of the human. This discussion of the body and soul in the years of 1846 is given in the sand of phrenology, a way that justifies the reason for this relationship of the rational soul in the body, duality is not only occupation of spirituality but materialist, clearly these materialists are not dialectical, are the germ of the positivists. Central ideas are the pursuit of thought and soul, inhabited in the brain, in some specific area of the corpus callosum or, conversely, are dissipated as a whole that does not reside anywhere. Discerning this discussion is important in deciding whether the religion has its preserve and other the science, in the body. Claims only allow the fragmentation to be a concept, a lifestyle, the message was: science and religion can get along if each takes care of its work. The result is that spirituality and organic seem to have different owners in terms of social participation, a body divided into two. But literature tells us that became more fragmentation and deformed bodies astonishment only show the intention to know the mysteries of this process was not regular but generational indicator of impairment by malnutrition.

The years 1846 to 1874, are indicators of the process which separates the science policy from positivists with the Escuela Nacional Preparatoria (National Preparatory School) and only recording of the press gives us so deep that discussion about the importance of making a new body, whether materialistic or religious. Positivist truths are assimilated, but are silent forms establishing the double discourse; there were liberal religious, believers. The simulation has a history.

The discussion about what is the importance and success of the best interpretation of the body, in the century we see, remains unanswered. There are a number of approaches on the discovery and the need for groups of knowledge about what it means pathologic and the normal; diversity of looks focus to the face, to the figure, the concepts of death and even how to punish a body, which is the criterion to identify parts of the body, reflexology philosophical, scientific view left between the ideological background of a dispute over the Conservatives and Liberals, tinged with the idea of seeking the truth, let out the stench of ideological persecution and provide for disqualification for the sake of common good, the political management that is beneficial to what is said by this or that author and especially is a foreign stay, the disqualification only minimizes the Mexican thinkers (from their story of dwarfs) from the stories of Motolinia in the Cologne[39]. So bring on the French model to classify the Indians and the poor is not just a whim or an anthropological curiosity, is the systematic cleansing of one group by another to erase the memory of a culture that is "nigger in rice or black sheep". Sort and identify are part of the same, in relation to the human body, so the details are so specific that is not enough to know who was Indian or poor, also their last names, their names will be an indicator for ranking and selecting, the repetition could be established stigma of violent behavior or to carry the seeds of discontent in his life. Multiple resources are to disqualify or segregate, not only the skin color is a complaint on its own, also features, the type of hair and clothing, a body dressed in a suit but with these characteristics may be fugitive of ignorance and the voluntary servility.

For all that I consider necessary to make some conclusions and I subtitled:

For a history of the human body in Mexico: reflections.

I

The possibilities of interpretation of the human body in the XXI century are avalanche of constructions wanting to realize various implications of a body in the present, in the past and its multiplicity of approaches to therapy, marketing, subliminal, the hormonal, sexuality, etc., that means that the body is seen as the body to make wealth, that reality may seem a truism, but I believe necessary to do a reading of the Mexican human body with the use of history, anthropology, politics and feeding, the cultural process and death. Understanding the complexity of this is only possible

---

[39] Colonial historians say that New World had small trees and awful land in production; human beings who inhabited it were short. To make things worse it was said that there was no possibility of a Spanish could come to study because there were no conditions to study and even worse there were no books. Available: Schneider, Luis Mario, *Ruptura y continuidad. La literatura mexicana en polémica.* Fondo de Cultura Económica, México, Colección Popular, 1986. p.200.

with a look to go to that social historical process that is responsive to the implication of actions that are not read with a critical eye, look natural and are a reality without history and that status is looking at the immediate response to the problems of historical order, so that responses are strange or only palliative not representing new realities of the corporal future, that is to say, alternative building processes are not actually a reality, they become a mechanism or a process of fictitious metalanguage that gives us the hope of viewing the new directions in building a healthier body and the aspiration of a dignified death.

The depth of this reality must unravel in the process of origin and how you will add new political parties to meet the intent of groups in the power of health, economic and political, the use of resources that can help the body of a racial group to be minimized, it is only designed for the purpose of domain, which means that the future of "that work" will not be anytime soon. Future generations will not unravel the thread of the story that asphyxiation, nullify and exclude from this political process, where the body and its appearance play an important role, defending the integrity is not given, unless the rights of their land, their sense of identity, many of the social injustices have that background.

II

History or historicizing the body of a large section of Mexicans brings out the nuances of construction in social groups. We find clear differences based on access to food, health and type of work being done. A body that is the item that allows you to read the past and the present must be read with the intersection of power and the wishes of the groups of political power. As the intention of maintaining that lifestyle ends in a pit or in a grave and starts a new one. I think it's a misunderstanding because a body has been subjected to violent processes from the standpoint of ideological, political, and nutritional epic new international cooperation processes, that is, unpublished networks to face the new circumstances, this involves a process that combines emotional condition, physical and spiritual, in close correlation with food systems. No doubt the inheritance to the children is not linear or one-one, is a process that is extensive, is nullified or create new forms of resistance.

The expression of this condition is made at any time of the individual's life, there is an election process where the subject opens new expressions and ways of feeling, but may be present in the subject according to their historical consciousness or, failing can get involved in new religious practices, nutritional, or cultural, that can open new horizons in their business of living and may even make new networks to enhance their own feelings. A body is not only the wishes of parents, is the historical concatenation of a network that managed to survive a fundamental principle of cooperation. Known is that the organisms most competitive are those dying sooner, therefore are the anti life of the future. So many Mexicans have survived; their ancestors were supportive and cooperative. When this spirit is changed in the city or in urban areas by competition or the destruction of others, are to gestate new possibilities that are destroying the lives of families or certain traditions of solidarity, even the death type changes, either age or shape, but could be very prolonged agony and fulminant. Then the competition to be the strongest, the bravest even better, carries a body full of suffering and sorrow develop as a solution to exist, making it possible to split the body in mind-body and nullifies the body as a process manager in social reality: think with your body as a whole becomes somewhat utopian and out of sense. The denial of the body as need to think, feel and act with him. It replaces the reason that seizes him and rule in their cooperation activities between organs and produces not only somatic disorders, if cellular changes, glandular secretions and even oncogenesis.

This situation is typical of our time and culture of competence. What makes reading difficult and generates a lot of smoke at the variety of symbols that can be taken for immediate readings of the body and its performance, according to the representations and meanings of any "scientific approach".

III

Therefore, consider going into the past and rebuild part of that process we have in the clinic (with patients) and ordinarily live without contact with their historical memory. We consider that the taking of Tenochtitlan by the Spanish in 1521, mark key aspects on the operation and construction of networks of cooperation of the body. Here are some aspects:

Bacteria's exchange. Strengthening the immune system of indigenous and Spanish.

The enrichment of diet and changes, syncretism food, including the incorporation of the cow, pig, sheep, goat, horse and donkey were and are the breeding ground for new processes combine health and gestation of new diseases. Add missing all industries that are derived from these animals and the earth changes, flora and fauna.

The regulation of health, as the regulator *protomedicato* (organization founded in the XVI century to regulate medical practice in Mexico.) about what and how to cure and above all who should do it. It is the official exclusion of a concept of body different from the Spanish. With the stigma of the demonic or ignorance could kill any culture.

The imposition of a religious belief that embodies in the body and allows the food of blame. Is the mechanism that has to produce more disorders in Mexico and its realization in the body. The culture of fear and guilt may minimize or even deformed body posture and its attitude. This disrupts the body's internal organs; an organ is able to dominate others via an emotion. The tyrant also occurs inside the body, could not be otherwise a tyrant.

The Inquisition was not only a violent way of putting fear in the body was also a constant censorship own and others' actions, and that stops the positive attitude, innovative. It takes a dose of rebellion to break this straitjacket that immobilizes and life threatening. Neutering can be a violent term but fits this idea.

The exhibition of the groups in power may be part of a role model and sets boundaries on individuals, some events may be existential crisis and frustration combined with the bitterness, gives an undesirable existence. The question who am I? No results in search of the true self, wants to be like the wealthy, then his true self does not appear, his wishes are those of others and he is diluted in a sense of suffering: that of having to deserve an idea becomes unreachable that only allows resignation to the fatalism of a manifest destiny in his body.

The health workers do not address the body. Their work proposals only address concepts, there is much discussion on the categories and their development, even as a concept is evolving, it may be the disease x, but not studied the process of building a body does not seem to matter.

A doctor, a psychiatrist, psychologist, scrutinize, standings, attitudes, behaviors, symptoms, pain etc., but the body does not appear, clear the possibility of understanding the process in the individual, looks how a disease behaves, a condition in the body, but this does not seem to be important. The institutions have been formed around this principle; systems and forms are standardized and related in language that has its counterpart in the actions and expressions of the talk of the problem in the body. So it is not important in this logic, the individual's body, the ego, his condition can feed what you want to study with him.

A man and his sufferings have two instances of the same fragmentation, first symptoms corresponding to the elaboration that represents a process gestated from observing how it evolves in the body, this means that the body only serves as the enabler of observation to systematize any disease or condition, which relegates the body as an element that can see the behavior of a bacterium or virus, another to know the human body requires a different logic, the combination of a historical-cultural process to know the building of a body that had no consciousness of being a choice, with that knowledge, it is possible to consider the possibility of elaborate the body at will and can defend itself better against the conviction of history, it approaches the work of individuality, to develop diagnoses according to the body of the subject even if it is the same disease.

I would like to close these reflections making clear this historiographical and historical approach shows a vicious circle and denies the possibility of alternatives of corporal existence, determinism has his life insured, on the contrary the intention is to make a circle in life; where existence becomes a commitment to biography and culture that has its history in order to plan the present. As regards.

3

THE SOCIAL CONSTRUCTION OF THE BODY IN THE XX CENTURY: THE ROUTE OF THE ORAL HISTORY

INTRODUCTION.

The XX century is considered the century of wars, but also the psychological problems, characterized by stress, suicide, depression, anxiety, dementia, and so on. Already in the early years -1900-1920 - is given a quest to explain the psychological disorders as well, we explore among other things, the five senses, sleep disorders, dementia, crime, prostitution and learning problems in children, concluding that social reality had a momentous impact in shaping people's bodies. Industrial society disrupted personal life and brought new health problems that in turn implied a change in the process of living the human body[40].

Scientific rationality faced a crisis in the psychological explanations, and just to say, the body was seen as an obstacle to interpretation. Anthropological studies, meanwhile, also sought to strengthen the importance of culture, ethnology devoting to the study of emotions in

---

[40]   Historical, social and economic studies show that industrial society transformed lifestyles of people and their relationship with nature; this separation between man and nature has its effects on the human body even today.

different cultures. The ethnographic work revealed several processes that demonstrated the diversity of buildings in the cultures[41].

The human body - and his study in different fields of knowledge-lived, since the beginning of the XX century, a quiet transformation with new diseases, and began to emerge a society which produced bodies with chronic degenerative diseases, premature deaths and in general, a quality of life unworthy. This fact reveals that "something" was happening in the postindustrial society and the body was the synthesis of this process, with different nuances depending on where they deal in the social division of labor, culture and economy, while still consider the body as the result of a family genealogy, in which the subject also makes a choice either consciously or unconsciously[42].

This new corporal reality was actually an example of the changes that the societies were suffering, with individuals increasingly their frail health, which had led to an average life of 68 years for the year 2000. The conception of Mexican society as a field where building a new and complex social process, will allow to see the body as a space in which two fundamental elements are interrelated; the organs and emotions. However, the imbalance of these elements not only brings implications for the health, if not how this corporal process is intended to be addressed, this means, with mono-causal theoretical explanations or the proposal of homeostatic cycle, on one hand, or the social determination of disease or holistic-comprehensive proposals, on the other.

Facing it requires a new reading of the body that goes a little beyond the purely causal, as well as a careful reading of the instruments we use

---

[41] For more information the interested reader could read: Salvador Castellote, *Compendio de antropología*, Valencia, Edicepi, 1999; Wiliam Y. Adams, *Las raíces filosóficas de la antropología*. Madrid, Trota, 2003; Tomas Barfield (ed). *Diccionario de antropología*, México, Siglo XXI; George Devos, *Antropología Psicológica*. Barcelona, Anagrama, 1981.

[42] You can deepen on the subject: Mario Camarena, *Jornaleros, tejedores y obreros. Historia social de los trabajadores textiles* de San Ángel, 1850-1930, México, Plaza y Valdés, 2001; Patrice van Eresel y Catherine Maillard, *Me pesan mis ancestros: la Psicogenealogía hoy*, México, CEAPAC Ediciones, 2004. Sergio López Ramos, *Lo corporal y lo psicosomático. Aproximaciones y Reflexiones*, 3 tomos, México, CEAPAC. Ediciones.

to decipher the new times in which we operate. Based on the above, we break down the following objectives: 1. To reconstruct the history of the body to understand the processes of emotion, and its implications, in the history of the XX century[43], 2. Drawing on oral history to identify the construction process of the body and emotional in chronic degenerative diseases[44]; and 3. To analyze this new physical reality to understand the history of our time with emerging conditions and consider the development of prevention and intervention strategies[45];

REDISCOVERING THE BEGINNING OF LIFE.

While the story is a science that generates controversy over interpretation of the facts, its merit is that it allows seeing the past to try to understand the present, with an optimistic outlook, can be helpful to plan the future. So, in this work uses this discipline to give a location to understanding the past that is manifested in the present, and hold its reflections on the processes we are interested in the Mexican society[46].

The social construction of the body in Mexico is part of a sociopolitical process that is also related to medicine and jurisprudence, the regulation of food systems and the establishment of hegemony over the concept of health, which involves a system of diagnosis and treatments.

---

[43] The process carried out is documented from the XIX and XX centuries. Available at Sergio López Ramos, *Fuentes hemerográficas para una historia del cuerpo humano en México*, México, CEAPAC Ediciones, 2005.

[44] The research we have done in the past five years has resulted in the volumes 4 and 5 of *lo Corporal y lo psicosomático. Aproximaciones y reflexiones*, México. 2008. There are several cases reported specifically on chronic degenerative diseases using history of patient's life.

[45] In the permanent research and updating seminar of the FES-Iztacala, "El cuerpo humano y sus vericuetos", have made the first attempts to develop a system of diagnosis and treatment for these conditions. Work that we hope will be completed by next year.

[46] For more information, see *Estadísticas históricas de México.* (tomos I y II, INEGI), which show the levels of production of raw materials and services, giving us an idea of possible correlations with some bodily effects.

In this sense, oral history is a means to reconstruct the life of a person, group or society, and then we can make a different interpretation of how an individual constructs his personal history and the origin of new disorders of health. Social history is in this sense, the expression of this process in the body, and oral history can be reconstructed, and for example, how the social body of a government is expressed in the body of the governed.

This process of building the body in individuals has three moments; social, family and personal. The family is a social condition that gives meaning to the significance that the subject makes of his personal history. The social construction of disease reveals how an individual constructs a body according to its time and space. This is important to understand how a particular disease is specified in the body. Society is the platform for the development of the body, so it is important to know how to give the individual's relationship within the same, with the understanding that we accept that an individual is the result of their culture and geography[47]. So it is not convincing to attribute absolute credibility to other theories or methods that have attempted to define the Mexican, as this reflects the space where it weaves his life and style[48].

The importance of an oral history that goes after the individual processes, family and community is revealed as an amalgam of ideas where the principle of life seems not to matter to the subjects. Thus, we find the expressions of a condition that undermines the subjects and submitted them to a lifestyle that condemns the continuation of a determinism, where it is possible to construct different options for living different in the same body and less of it as the space that can make sense of the implications of a world that conditions. Identify and unravel the conditionality reveals that lifestyles have become a stigma that makes sense of a body that

---

[47] For the theme of oral history can be found at Jorge E. Aceves (coord.) *Historia oral. Ensayos y aportes de investigación*, México, CIESAS, 2000 y del mismo autor *Historia oral e historia de vida. Teoría, métodos y técnicas. Una bibliografía comentada*, México, CIESAS, 1996.

[48] Extrapolation about readings of Mexicans is common and always ending in disqualifications. Early in 1901 Ezequiel A. Chávez Lavista, said in his essay *Rasgos distintivos del carácter mexicano*, there are five types of Mexicans and that no theory could be extrapolated to study, before making adjustments to adopt it to the Mexican society.

endorses contradiction to the establishment. But, what can we do for a body to regain its connection with the principle of life? First, to see where the principle of cooperation was lost or where it became a principle of survival and biological responses and behaviors were built that deviate from the principle of life[49]. Second, do therapeutic work with the body to recover the internal and external processes, and so find personal processes that obstruct the construction of the subject.

Our time is marked by emotions. The question is: How can we see the emotional body without losing the organic body? Where live the emotions and make a complex process in the body, altering the energy economy of individuals? Body movement cannot ignore the emotional processes.

An emotion today may become the consumer of energy in the body and get it over with the life of the person. Just know that death rates in people under 60 have increased significantly, this as an indication that the body is consuming more energy in periods getting shorter. Stress and anxiety transform into the interior rhythm of the body, and once changed this is not so easy to regulate. This emotional energy imbalance is present in the story of the ill persons studied[50].

Excessive consumption of energy in the body is one indicator that can be observed in a body that constantly fight in a highly competitive society. The person experiencing wear processes that arise when we do the story of his life, personal history with the family, their diet and their relationship to work.

The history of the body through oral history is a path that opens an in-depth picture of the individual, especially when making correlations

---

[49] Available at Fritjof Capra, *La trama de la vida*. Barcelona Anagrama 2000. For him, cooperation is a fundamental principle in prolonging life, even considering that this principle is what gives meaning to life and not competition.

[50] Chronic fatigue, exhaustion, collateral disturbances such as insomnia, sexual impotence, slow digestion and diarrhea often include many symptoms, expressions of a living body a reality that does not fit. And this is a breeding ground where they are generated chronic degenerative diseases. At Facultad de Estudios Superiores Iztacala have made over 30 thesis based on interviews with people with such problems. We can find diabetes, colitis, depression, infertility, gastritis, and so on.

with the pains, tastes, smells, emotions and ways of becoming addicted to a condition, a way of seeing or feeling. The reflection is deeper even as we speak of a principle to find routes that can take an emotional process that has implications for the organs. The task is to conceptualize the body as a space where the principle of life is the priority and not ideological or economic processes only, pay attention to bodily processes and find ways to build unity of the relationship of organs and emotions. This means that our look is not monocausal or multicausal, are more illustrative for our case, talk about microcosm that is constructed in relation to a macrocosm; this means, the body as a result of this micro-macro process. This is where we find the chronic degenerative processes of contemporary bodies.

SOMETIMES THE BEAUTY NOT REPRESENTS THE LANGUAGE.

"Flowers are beautiful because they do not speak." We can say the babies are beautiful until they start talking. Language is a resource that distorts human beings, as lies at the level of representation and not let the body work with free will[51], there is an articulation and linkage of language to bodily functions; the anger expressed in whatever form, may be an example: the secretion of bile juices alters the lining of the stomach.

Desire is an important point when the baby says "I want" begins to fall prey to the desires and suffers from the unfulfilled ones then the body is deformed in postures and gestures, and create a memory that will relate to the messages that are not met. The man or baby living prey of desires has to suffer the consequences. This builds the principle of conditionality establishing in the body and make the routes of emotion, turned into symptoms. The explanations of this process ponder the physiological aspect of the symptom, not linked to the process of emotions. Moreover, the metaphor of language can deceive subjects and rational thought becomes the resource that makes the subject apprehensive and possessive.

---

[51] I mean the construction process that is done with language, which is always related to the symbols and meanings that express a moral and ethics, so the language is not free from a culture and style can be incarnated in living and feeling the body.

Be beautiful in our days may be synonymous of not talk much, which would mean that existence is a contradiction, as there was no consistency between what you think and what you do. Talking is not a problem; it is a body part that expresses the cultural reality of their time, which involves the use of reason, thought and language. How does the body process this complex relationship? With the contradictions and adaptations to changing social reality. This process is where the body is faced with the choice to succumb or contract to the problems of his time to search for alternatives that, apart from dealing with this relationship to language, to build other relationships to find a solution to a style unnatural life, artificial: it is a foreign body in society today. Their language cannot represent the complexity of a highly competitive society, thus the person must try to unravel the reason, thought, language, conjugated with the process of family history in the body[52].

The problem with contemporary health concepts is that they no longer have meaning, however, the way they were made, sustained fragmentation, today still retain the same explanation of the signs and symptoms of the human body. Also important is the process of reflection that is built around the body that is involved in the dispute of thought, thinking the body with the slogan of the past is something that forces us to rethink the lifestyle of a deeply unequal society, and that created multiple processes to seek living with meaning:

---

[52]   With the methods that count today cannot understand the fragmentation of new symptoms manifested in the body. So the category of psychosomatic may be incomplete to decipher this bodily process that must be a conception that includes cultural and emotional. The above makes sense if we understand that a body is not different to the status of science and technology, designed to make it the same as hundred years ago: the body that has a headset on the ear, that feels the adrenaline as experience different from the ordinary, and only alters its construction process to live a decent life and longevity. If this reading is correct we may assume that it restricts freedom, in terms of education, is what kills the body and consequently does not cultivate the spirit. A body subject to the standard process and the rule necessarily has to stop being itself and become the bearer of a culture of reproduction, losing the spirit of new processes.

Every *manifestation* of anger, to be different from the anger he feels, already contains a reflection on him, and this reflection gives the emotion as individualized form phenomena proper to the outer surface. To express anger itself is a form of self-presentation. I decide it is appropriate for the appearance. In other words, the emotions they feel are so *intended* to demonstrate in its unaltered state, as are the internal organs through which we live. It is true that never could transform the appearance if not incite themselves to it, and if I do not feel like other sensations that make me aware of the internal life processes. But the way in which they arise without the intervention of reflection and the transfer of speech - a look, a gesture, an inarticulate sound - not unlike the way that higher animals communicate with each other, to human beings, very similar emotions[53].

We cannot accept that emotions are willing to demonstrate, and not from the standpoint of biological logic, as expressed according to the principle of life on the ontogeny and phylogeny.

But an emotion is expressed not only as the sense of a whole, a set that relates to a body, some organs; the emotion is expressed as part of this process, there is no division, no exclusion, what happens to a subject has a travel route inside, and not only is the fruit of a relationship with the body, is the result of a relationship that speaks of the interior and exterior. We can say that the interior, or emotion, is to transform the face and posture; also the relation of one organ to another, breaking the principle of cooperation for competition in the body, is the loss of a vital principle[54]:

But our soul experiences are so closely linked to the body to speak of "inner life" of the soul is as little like talking about a metaphorical

---

[53] Hannah Arendt, *La vida del espíritu*. Barcelona, Paidós, 2002, pp.55-56.
[54] While emotions are integral to living beings, their study from philosophy relegates that discussion of inside and outside, which puts us in a predicament if we take this to the psychological. If we accept that move and transform the interior, then there may be only the expression of a shared thing in the body.

inner sense by which people has sharp sensations of better or worse organ function. Clearly, a creature without spirit could not experience something like to an experience of personal identity, is totally at the mercy of his inner life process his feelings and emotions, whose continuos change does not differ in any from the continuos change of bodily organs. Each emotion is a somatic experience, my heart hurts when I'm sad; it heats up the sympathy, opens in the rare moments when love and joy overflow, and similar physical sensations overtake me as anger, rage, envy and other emotions[55].

With a reading like this, it feels the risk to accept that the duality is real hard to beat, give life to emotions to take hold of one, leaves us at the mercy of their whims and sudden changes. Emotions and their expression have a different route to take over the body. It seems that swarm and take over individuals. This misinterpretation leads to a discussion today makes no sense: if something takes hold of us or are just a condition of living beings, especially humans. The important thing is that its construction and development is done in a body that is the result of a society, in a specific culture[56].

Let us not stop to discuss whether the feelings are a reflex action or are part of the body in a construction of a time and space. That goes beyond reading, as a body cannot be just an expression of the organs.

If the inner psychic foundation of our personal appearance was not always the same, there could be no science of psychology, is that while science is based on a principle "inside we are all equal" just as physiology and medicine are based on the identity of the internal organs. Psychology, deep or psychoanalysis not discovers more than the ever-changing moods, the ups and downs of psychic life, and their results and conclusions are neither particularly attractive nor very significant in themselves[57].

---

[55] Hannah Arendt, *La vida del...op...cit.*, pp.56-57.

[56] Hannah Arendt says, "In other words, no sensations that correspond to mental activities and feelings of the psyche, the soul, are feelings we experience with bodily organs" (ibid, p.58). I agree, but necessarily have to chronicle that process.

[57] Ibid., 59

It is a physical reality that can not be divided to make sense of functionality, this means, to see the principle of fragmentation, where the body becomes an obstacle to the construction of an explanation that has unity with the emotions, less as a junction point with the culture and daily life where it generates the somatic[58].

Of course, the meaning of life can be difficult to locate, but we understand that life is on the bodies and their multiple relationships that allow them to build networks of internal and external cooperation. While there is no intention on the subject, it has a will that moves with the process of life inside of organs and emotions. It is our corporal world the space where you can better understand life; when is lost touch with human nature, forgetting the body process is enormous, and death is natural, because nothing surprises us, and we all living beings are like that[59].

Anyone can ask: Who am I? But the question should not go to another to tell me who I am; you must build, look for the answer in the body, meeting the country's history, culture, and family history. This will allow to understand the person as his body is built, very different to his choice and there begin the process of build; change, transform the personal process, that is the transformation of body memory, emotional. Coping with the instituted in the body will reveal suffering for desires and attachments, and conditionality that created the subject in his memory of life.

And this is where many remain stuck with a fact or an event, a claim, a hatred, a grudge, and so therefore not easy to make the jump, the impediment is big and understand it with the reason is no guarantee of change, and that cognition is a moment where you can see the problem, but not change it. This is the time it takes about a new body pedagogy to enable the person to express their internal states for the flow of energy, emotion. If this is not achieved, the travel path of the energy will be

---

58 This dissipation of the body as historical sentencing gives us the possibility to understand that emotions are not an issue of dispute over the body, or the opposition of thinking or learning how to feel.

59 Said San Agustín de Hipona: "... know therefore that there cannot be time without creature" (confesiones, Alianza, 1997). Of course, historical time is what allows the process of building the creature, and makes sense of the body building process.

hampered, which provides insight into the process of compliance, it can be turned if it takes action, the inner movement, and the flow of energy. When blocked, the subjects see the world in a way not real, imposed by culture.

The work of body is a process that claims to do awareness of the implications of having a body brought into disrepair, into oblivion. Knowing this forces us to seek other methodologies that can help people to develop new relationships with their bodies, without pain and with dignity, and with a perspective that goes beyond what instituted by culture. It is feasible to construct new way of being in the body with the body, the feasibility is a historical result, this means, they did other aspects but isolated, with specialization in certain parts of the body, its control can be a key for stated herein.

This process of body construction has to fit into an individual story, where the subject can "see", "meet and decide what to do with his body." This confronts us with the problem that each individual must know the implications that come with family and social construction to make his choice a fully informed. The depth of this issue leads to processes that result from this new condition of conceiving the body as a space for the construction of the body itself. Thus, if we stick to reason well we understand and explain what happens in the body and how is reinstituted as disease processes, rationally this is easy to understand, you can even make very clear explanations, but there is no contact with the inner process, so it is not feasible beyond reason. This is a kind of self-deception, for the body to get excited and spend more energy. From this perspective strengthens individualism and could not build more than illusions: knows everything, understands everything but does not move. Standing still is not appropriate at this attempt to build a different body; the movement is very important as it will prevent the energy to stay somewhere in the body and start doing damage to organs. Without movement the emotions are conditioned, because do not circulate. Chronic degenerative diseases are the result of this passivity and compliance of the organs, which means the establishment of a body memory that determines the kind of inner movement, being recorded as normal in the individual. This behavior will be seen as part of a lifestyle, "so is the person," and will remove the possibility of being the same.

If we understand that a body is also constructed with the ideas of others, we can see how it mimics the process, since individuals are easy prey to become a world full of corporal fantasies. You can say "I change when I want" but the reality is that this is not so easy. The point is that representation of the idea that represents the body as having the control. An example of this is addicts, they do not have the willpower to put a spin on their actions, and the word cannot alone explain the reality of the body. This is where the idea makes sense to work with the body, because, as stated above, it is necessary to move the body energy and avoid stagnation at some body part of the individual.

However, it is important to mention that sometimes this change is only an illusion, as though the person, cry, yell, hit, curse and even vomit and convulse, this may be just a deception of the individual, constructs the idea of having changed when in fact nothing has happened transcendent in his body. The entrance of a reflection changes a little voice; it is a tender look, and so on. But nothing happens, the change in memory does not come, and this is when you wonder if the corporal and emotional have been conditioned, because it is difficult to change to a job that does not include body therapy.

The body of XX century forms emotional expressions that produce these health disorders. The challenge is to change the memory to provide better living opportunities for people with disease, especially called chronic degenerative.

4

URBAN BODY IN BUILDING PROCESS: AN APPROACH FROM THE EMOTIONAL AND THE ORGANIC

INTRODUCTION.

Do Mexicans are special? The answer is no. We are human beings result of a heterogeneous culture that has its genetic roots in Asia; the mixture with the Spanish made a mixed race or multicultural. In these times where the purity of race has become an ideological or economic resource is not feasible to attempt to explain the new reality of Mexicans with easy assertions or denials of the times. Exit the geographical area where we gave birth, open the interpretive horizon to new way of living with others: build inside the body with new possibilities with a working concept that goes beyond the mechanical idea of joints confronts us with the reconceptualization of new body processes that encourage us to seek new routes in the body construction, how has been built this way of life that binds us, consume us, damage our inside, it allows us to learn the living process in space and time while we live. Build in the geographic space that we were born is a minimum principle of congruence to transform the immediate reality of an individual; otherwise the process becomes fantasies, hopes, thwarted desires of greatness and fails the realization of alternatives.

To make such an attempt to work with the body of Mexicans need to go to the concepts in which their body is built, because that is where containment processes are established, this is, establishing bodily and psychologically how to feel or living the body, the forms instituted as truth about the functioning of the body become a dogma or a reality that can not be challenged or changed, that implies work of a conception of physical reality that subscribes to a mechanism for the regulation of body expected to act and work in a certain way, it means a normal historical process that fits a way to make it subject to policy or morality, a religion even to any activity you want. Discovering this process through the story and follow its route on time, not allowing to reconstruct how a body has been generationally becoming malnourished or small, without will, without possibilities of building options in his present, the culture of fear generations becomes meat generations and freezes it, that makes no other perspectives on what it means to live or feel with the body. The heaviness of a culture of oppression does not give opportunity to see what you want to be or wants to stop being, learning to see and feel the body's process is not something the illuminati or the work they do to get rid of stress or the oppression of the body that is exploited in any area of the neoliberal consumer societies of our time, the approach is to go deeper to the ways of making choices in our times of chronic economic crisis.

Therefore, speaking of the body of Mexicans is not just a health discourse, no; we talk about the ways of how to preserve life or just finish with it in many ways, established or with those derived from the initiatives occur in the need not to lose its life for individuals. The confusion, despair, anarchy, violence, etc., Can be part of that process that allows individuals to realize their life plan. Where to go or where to begin to propose new visions of what it means to build in the immediate area which is the body.

Routes can be many, but feasible is exercised to the reconstruction of how a body is done historically, one level is the conceptual and the other is the matter, the realization of the ways to impose lifestyles, eating certain food, marriage or death are among the indicators that can be considered to give a fresh reading of the times we have lived. In this logic the concepts of body and emotion in force in allopathic medicine allow us to understand how an incarnated category may hamper the process of life itself and becomes a shared socialization or a kind of complicity on how to operate inside the human body through the stigma of an organ or an emotion. The problem is compounded when we are laying down in ways that are convictions on the physiological, endocrine, or emotional, involving the denial of new possibilities of reading the body or life itself. It is not a mess, not a puzzle difficult to solve. Is required to travel the route of these categories and how to make a lifestyle that truly becomes the body maps are as geographic: we have alternated for establishing that Europe is bigger than Latin America and thus establishing a psychological sense of power because it is larger. With the body maps become a reality that is flatter and seems to stand still, of course that the concept of body is dominant, nobody tells us it is a map that can change its operation by building processes that is the subject in their culture or that relates to its character as a person who decides to do something with his body, whatever, a cancer, a chronic or excessive care that leads to death, in the worst individuals do not know what to do with the body and vomit, do not feed it, punched it, led it to states that are not the most balanced, either with drugs or sex, and even work as a single resource, the point to know that is most celebrated, most significantly for the space of his family or he wants to make others feel using his body as an instrument of notice of death or abuse, how the media distorts an individual with his body are to the *nth* power, example, it is not clear what you want to build with the body here and now, is not feasible to make sense of a body that is represented by a map that can never be read in conjunction with a space and time, subjects rarely know where are the guts. And that indicates the distance that people have with their body, it is fair to say that education should never posed pedagogy or education body, the cognitive weighting creates a distance to learn with and from the body. It means living the great majority of his life with ideas outside the body, neglect is a condition of life, and until the organs are in crisis, there is destruction or disease when the subject can recognize his body map and become expert, in other cases death has chance to pick him up.

Learn how to conceptualize an organ, an emotion is an important point for our work, hence deriving the ways of how individuals can live their body or take away from their condition. The fragmentation of knowledge established the individual becomes someone who can not know everything, whom should not approach a space as intimate as the body and that exploits an attitude of fear or knowledge becomes more complex than an organ or emotion, the duality that has cultivated the body of scientific rationality, it does not let you to know how a subject is articulated with his body and deeds processes or overexploited an organ or viscera, the body has its timing to action as a lifestyle that is located in the space you live, this implies a reading that allows us to make sense of a body in the present, so a read emotion outside the body can mean endless readings in individuals, leading to confusion and asserts that a thing is the emotional and another the organic and it becomes corporal truth, to live splitted is a style, there is imbalance in the interior and exterior of the body, emotional and organic harmony is something distant and unattainable.

Where to read this organic and emotional process? 1) A basic concept is the unit body-emotion and existence in space-time of a culture. Because only then can we understand the behavior of a body in free route links with social, cultural, and construction within the family history. 2) Conceptualize the relationship body-emotion as part of a social history that allows the subject to have an election with a flavor, an emotion, and that brings us to how to build an organic constitution with certain conditions, to identify whether your problem is organic or emotional. 3) An emotion is as vital as any other body part, think otherwise is to maul the human condition and incomplete approximations to the corporal process, this means you can not understand the liver, without anger or kidney without fear, emotions that make the subject is able to preserve his life or kill it and reverse it the same with any other body or emotion.

4) Recognize that the body becomes a body process that establishes the principle of specific historical moment of a society. The depth of this reading will agree with the work process to be taken by the individual, have control of desires or overcome any hatred or resentment. 5) The process of an organ can not only be understood with the readings of imaging studies, it is also feasible to make sense to know to find out the extensive network of the organ in the visible parts of the body and that is where we understand that a body has a chain that makes it difficult to prevent something exists independently. With this reading of the body as space is concatenated and gives meaning to a long kind of inner life and with that you can understand the depth of a body can not have

social autonomy, or inside of the organs and emotions 6) A disease is the expression of protest of a body, because it always tells, here you have to wonder how or why we do not listen, what to do to listen and make sense of a beginning as life and not allow that go missing every day.

Some clues in the XIX century

The XIX century Mexican society faced the problem of the plague epidemics, pollution, unsafe, adulterated food, housing poorly designed or overcrowding, explained the phenomenon of death not only requires statistical data or a sequence. The validity and credibility is a very interesting point that requires careful reading to avoid making statements and is not easy to believe what the data says. Find the correlation with other sites is to demystify the process or not to condemn them all to a difficult practice of death. The regularity of fact, it may be a truth that should not be denied or questioned. In 1886 Antonio Peñafiel his contribution to this point and wrote an article: "Statistics Applications to medical science[60]." It explains the uses of statistics in the health field, illustrates the problems of health from 1869 to 1878, where death rates can be identified by amounts ranging from 7 to 10 thousand deaths per year in an average city of 350 thousand inhabitants. Reconsider allowing the figure because that amount had to do with correlation of birth and to know whether the measure population remains. Search all data required for a conception of them, Peñafiel said:

1. - The *data* must be true, exact: Without this condition may have no value other statistical operations that they practice: a bronze statue, does not stand solidly on clay pedestal.
2. - The non-accurate data cannot be accepted except with the category of *possible or probable*, as appropriate.

---

[60] Antonio Peñafiel "Aplicaciones de las estadísticas a las ciencias médicas" en *Gaceta médica de México*. Tomo XXI número 2 México, 15 de enero de 1886, p. 25-33. The author devoted to the study of national population statistics, made calculations on the inhabitants of the country and Mexico City.

3. - To practice with caution and circumspection the investigations made in order to obtain data.
4. - That the source of the *data* has clean patent: there is no interest in changing or the essence or the number, or quantity, or the value they represent.
5. - That all scientific research gathers abundance of data, this is, to reach large numbers.
6. - Probable data do not interfere with absolute consequences[61].

This proposal invites to see the statistics as a resource that can enable us to find the laws governing a phenomenon. The discussion on the validity and its use is not exempt from the proposal that used to legitimize the readings that serve to justify political decisions, Peñafiel going to health and said:

> The statistical operations only provide us with the *laws* of phenomena. How to get to the causes? How after the constant collection of well-observed facts rather homogeneous data added, large numbers considered, we will reach the truth, the truth causes and explanations of the clinical facts of the science of healing, of medicine in the end[62]?

To illustrate the above gives examples of the causes of the disease that determines death and said on the subject:

> *Site* in which we live, the *air* we breathe, food they use and does use bridging to the present case, to proceed by synthesis, to determine the *influence* they can have, the height we live up, the polluted air and poorly oxidized we breathe in Mexico City, the insufficient and bad habits and worse drinking water, used by residents of the capital. The topographic situation of the city and its height *predispose* to

---

[61]   Antonio Peñafiel, op. Cit., p.29
[62]   Ibid, p.30.

general thoracic diseases, heart, heart and lungs, low-nutrient foods and unsafe water determine gastrointestinal diseases[63].

It is an explanation that can be on top of their time. Locate the subject in a process geography and lifestyle that makes a disease condition, only allows us to approach what we hold in the process of health, the involvement of emotions in the construction of the disease by subject. Peñafiel claims that the air and food and height affect the lungs, heart and intestines. These truths are adjusted to the circumstances and the statistics for the years 1861-1881 the percentages were 28% of respiratory tract disease and 20% of gastrointestinal complications. His explanation is correct. Find the relationship with the space where the disease occurs. Perhaps the temporal relationship is not a constant. However it seems to be a constant throughout the process of Health in Mexico City. Statistics from the centuries XVI, XVII, XIII, XIX and XX show that these health problems are kept and maintained as part of a more complex problem that not only relates to air, food and contaminated water. Antonio Penafiel quotes Quetelet saying:

If the identity rigorously exist in all men, suffice a single disease well enough observed and followed by healing, to obtain the same result, provided that the same disease occur in other individuals, but this perfect identity may never exist, you must believe so, considering the diversity of ages, sexes and constitutions, past illnesses and countless other causes. A physician throughout the course of his life will not work maybe twice in circumstances quite similar [64].

While it is true to say, that could not be considered as a working principle that the process of building the individual is unique. The choice according to his circumstance allows his to build a body with a unique health status. Can hardly find two persons having the same process of developing the disease. The statistics bring us to the truth of the disease. The problem is how each individual can interpret the fact, and not only

63   Ibid.
64   Antonio Peñafiel, *op cit.*, p33.

that, which is to explain according to their scheme, for that matter, is that health becomes the principle of reflection and Peñafiel, whom without being doctor holds three aspects to consider as air, food and water; factors that affect the health process: it is a valuable contribution. It looked like things to consider. If we see this process as an alternative to the status of an epidemiology looking at bacteria, it is purposeful to have a new reading of this process. In the distance we can say that a constant in lungs and large intestine show the dominant disease, only leads us to think of another explanatory route, as is the possibility of the emotional. Sadness as the dominant principle of a body is being weakened; it becomes autoimmune and is easy prey for opportunistic diseases. This explanation can be sustained on a conception of body as a unit, just what Peñafiel said, geography, but the human being is built not only in geography, we must go to the social memory that builds types of felt emotional and sentimentally. The family space plays a substantial process with the genealogy of the lifestyle, the culture of the invasion, violence and fear or stress, you have to give a new meaning to the body of an individual who has to build new responses to adapt to the social relations becoming more complex symbols, meanings and representations. The exaltation of cognition puts away from the alternative body building process and approaches to conditionality.

By 1899 Dr. G. Mendizábal brings his "Contribution to the study of the flu in Mexico," says that since the first raid that made the flu in the state of Veracruz in the month of February in year of 1890[65] "caused great havoc in that city. Although Pfeiffer's bacillus, which was reported as the causative agent, referrals that result in death, plots not directly attributed to the flu are side effects, Mendizábal said:

> That flu alone rarely causes death, but because of what caused it, which is the same. Record mortality statistics for the time you will see that pneumonia, acute and chronic bronchitis, laryngitis, pleurisy, lung congestion, asthma, pulmonary tuberculosis, and in general

---

[65] G. Mendizábal, "Contribución al estudio de la gripa en México." In Gaceta médica. *Periódico de la academia nacional de medicina de México.* Agosto 1 de 1899, México, pp. 349-361.

all the affections of the organs breathing, give two or three times a contingent of mortality, and these diseases are engendered some, but all compounded, no doubt, by the flu[66].

The cause is the flu and its impact in meningitis suggests that its effects are implications of death in the vast majority of the population. The relationship with meningitis occurs with the temperature effects that damage the meninges and cause death. The table Mendizábal realized by José Olvera gives an idea of rates from 1894 to 1898 where meningitis is the simplest cause of 398 deaths to 505.

These figures give an indication of the effects of flu and its difficulty to control body temperature. The drug ingested was chinin-hydrochlorid. His reflection leads him to consider the impact and status of a patient and says:

> Depression is so marked in the physical, intellectual and moral order. Usually happens in all diseases that each patient is convalescing in right relationship with the conditions of his temperament and his previous pathological state, this lady was suffering several days from trigeminal and intercostal neuralgias and all consequences that come with malnutrition of the nervous system[67].

His observation is correct the particularity of the patient can make sense of his explanation. We found that the depressive state allows to understand the construction process that relates to the pathological picture can be understood at two moments one before the flu and another after worsen the deterioration of the body. This experience becomes an important principle to understand the body's responses to an epidemic like flu. True, there is evidence of a depressed condition of the population, but also those who migrate from rural to urban areas are more vulnerable to a virus or a bacterium that does not exist in memory diseases. So not only flu is a disease in isolation, is part of a social microcosm principle, geographical, the port of Veracruz and Mexico City. The statistics read

---

[66] G. Mendizábal, op. Cit. P.351
[67] Ibid., P. 357.

from the claims of a positivistic philosophy want to search for causes and Mendizábal dares to consider a new element as the emotional, intellectual and moral state, but is far from finding new body processes. It is a good indicator that reality imposed, explanation could not be free from the dominant medical conception.

Health readings vary and change. But maintaining the readings of a conception, which is not explained at all. Dr. José Olvera argued that:

> Amid the darkness in which science is about the functioning of the bodies to whom belong the psychic functions presumably most likely of certainty that in the regions of the brain devoted to the fulfillment of these functions, there must be cellular elements obliged to work under nervous excitement they get from a uniformly intimate relationship with their special structure, a structure that should vary in each class but can not distinguish the difference means that now exist, it is also true that despite the mystery in which nature has hidden admirable properties of each of the elements dedicated to the psychic functions sometimes fleetingly, the more prolonged and indefinite time, as evidenced by the faculty of memory, probably stemming from those differences or the strength of the impression or duration of action´s cause that impresses all this resulting from oblivion the memory sooner or lasting or enduring engraving in the organs of memory, but it is necessary to note that the quality of certainly this ability depends mainly on the number and perfection of structure of the elements of the organ[68].

Approaching the beginning of the relationship of the organs and brain. Find out the reasons for suicide potential relationships, words underpin a reflection that transcends the physical problem. Find other elements in society that allow you to find the trip of the social and organic. The construction is done in and by the individual finds its point of intersection with;

---

[68] José, Olvera, "Algunas palabras sobre el suicidio" In Gaceta Médica, *periódico de la academia nacional de medicina de México.* Tomo XXXVI, Número 19, octubre 1 de 1899, pp. 474-485.

The good moral *ingestion* will produce public moral health this is, that the criminals are also in the minority. If moral hygiene is helped in other ways that strengthen the spirit, this will approach the sublime ideal of supreme happiness. Unlike in case, the denial, skepticism, in short, injured with rough chisel the nerve cells, psychological functions belong to the pathology and physiology not moral and it is because then the impressions are very efficient in producing causes malignancies of the moral sense and as the action continues into the morbid causes on the body, determines chronic disease states, the more rebellious to medicine the smaller the resistance that opposes the power nature of the causes, if work indefinitely, disease is inveterate, and the evil, morbid state of mind, is a chronic disease that is almost always inveterate and irremediable[69].

His relationship with morality, with the process of nerves and pathology related to the state of health of the body, but the moral physiology as -Olvera- has an impact on disorders of the body. Make a bridge in the construction process that gives meaning to the disease that transcends the body and is expressed in the spirit. The author approaches the process that is socially in the body, the relationship with the body's process that he calls moral hygiene, constitutes a substantial element to understand the complexity of the relationship is with the emotional, organic and cultural. This relationship culminates in illness or in his absence the highest ideal of happiness poses the beginning of a choice of subject. But the consideration of moral hygiene gives us another reading of the chronic degenerative process linked by Olvera with the morbid and logic allows him to understand that there is a relationship between social and organic, but how to find the process of happiness, preventing suicide, and chronic diseases. It's 1899. Olvera finds an outlet to explain the suicide, attributed to the influence of the novel and the tragedy that exposed models, shifts its reflection on the use of red red in the press. But his approach to suicide and chronic diseases, show that the decline of health systems and public health is faced with the dilemma of denial within it and look for causes in the space of a novel and tragedy. Even so, one can see that the chronic

---

[69]   José, Olvera,, *Op. Cit.*, pp.480-481.

is not something different to its historic condition, Porfirian society was in the process of entering the XX century and industrial societies have allowed the body to build a new organic status and emotional changes, which may explain the emotional and organizational changes fail to articulate our author with the cultural process of Porfirian society, that is, for his romantic moment, hence the assertion of the influence of the novel and the tragedy.

Porfírio Parra was right in his article "Psychology of medical science", stating that none of the works of medicine did justice to the working method and subjectivity. He said to this matter:

> As if the methodology and psychology of science points were so despicable and so known to all that it was a waste of time to treat them! Deep and regrettable forgotten that, in our humble opinion, vitiates the contemporary medical education, and we are sure we will disappear as well as the trend grows in great spirits and the course starts to print their research[70]!

The forecast failed to Porfírio Parra, because to date the state of science has not changed enough to work on this aspect of teaching, which becomes an obstacle to knowing the human body, the learning process not only is memorization as well seen by Dr. Parra, need to consider another construction process that is done in the work of teaching, the multiplicity of relationships that are specified, they crystallize in an event such as surgery and not only an act of memory, Parra makes a speech that displayed a more complex process and concludes:

> Please note that medical science because of its size and variety is a real microcosm, in which there is scope for all abilities to be exercised and position to all vocations[71].

---

[70] Porfírio Parra, "Psicología de las ciencias médicas" In *Gaceta de México. Periódico de la academia nacional de medicina de México.* Tomo XXXIII, enero de 1896 número 1, pp. 579-600.

[71] Porfírio Parra, op. Cit., pp.579-600.

He was saying that we must not close our eyes to new challenges and demands in the training of doctors: Very successful his opinion surely had before his eyes, in the vocation of teacher who exercised, the loss of sensitivity to the formation of future doctors, central point in the practice of medicine. Not only was this voice, and Dr. M. Soriano could see in his "History of the National Medicine. Some notes on the proto-medicato[72]." It was necessary to review the formation and regulation of physicians, surgeons, apothecaries, barbers, dentists, *hernistas* (people who used to cure or treat herniated people), opticians, algebraists and midwives. Lost the relationship with the healthy and the sick body, and especially its method of cure. The discussion continues to this day it seems that the timing is something of days rather than decades. And the problem is resolved. Statistics showed these social changes and processes showing that it is not feasible to decipher only a reading of numbers and blame in the process of explanation of viruses and bacteria. The reports were made of persons killed per month; year and their correlation with places only showed that there was no control of epidemics and chronic diseases. The rates are expressed more in children, recurrent infections, viral diseases prevalent from 1879 to 1880; the difference is +258 in the population of 1 to 12 years[73]. It is noteworthy that only certain organs are most affected. For lung death figures were:

Bronchitis	215
Capillary bronchitis	147

---

[72] M. Soriano, "Historia de la medicina nacional algunos apuntes sobre el proto-medicato" In *Gaceta médica México. Periódico de la academia nacional de medicina de México.* Tomo XXVI, número 22, noviembre 15 de 1899, pp. 563-580.

[73] Although one can argue that there is another factor and the consumption of cow's milk contains many contaminants. "Can cause a number of infectious diseases: tuberculosis, typhoid, infant cholera, dysentery, diphtheria, coal mouth disease, actinomycosis, measles, scarlet fever, Asiatic cholera and plague, the most constant and terrible of these diseases, tuberculosis, "p, 464, J. E. Monjaras, "La leche de vaca y sus relaciones con la higiene pública" In *Gaceta médica de México. Periódico de la academia nacional de medicina de México.* Tomo IX, tercera serie, núm. 1 enero de 1914, pp. 463-468.

Whooping cough	88
Pulmonary congestion	123
Pneumonia	1550
Tuberculosis	680
Emphysema	105
For the heart the figures are: Heart disease.	322
For intestine and stomach: Gastritis	15
Stomach cancer	28
Gastroenteritis	74
Enteritis with diarrhea	1122
Enterocolitis	771
Colitis with dysentery	237
Peritonitis	129
Typhus	186
For liver the figures were as follows: Hepatitis	148
Liver abscess	38
Biliousness	179
Cervical cancer is expressed as:	61[74]

Intestinal disorders are the largest index reflect and are followed by lung, are the two organs most affected by social policies. It should be a reflective approach on the incidence in these two organs. May be multiple explanations and all are logical and compelling: 1) Poor health in the water and the air is possible to understand that they have predominantly gastrointestinal, as well as respiratory. 2) Although we try to make connection with the milk and lungs for understanding tuberculosis and increasing problems of the lower respiratory tract such as pneumonia, combined with changes in temperature and low immune system due to depression or sadness that the time had its period of crisis, low fruit and

[74] "Resumen general de la mortalidad habida en la ciudad de México en el año de 1879" In Boletín del consejo superior de salubridad del D. F. Julio de 1880 a junio de 1881, tomo I, 1881, pp. 141-142.

vegetable intake allows the body to have more acidity and produce more mucus membranes. Food in Mexico City and other cities was not rich in its variety, the problems of the alteration of food was a constant not take into account that nutrients were correct, altering the chocolate, milk, bread, coffee, wine, even the viscera were consumed or meat was horse or donkey. So the combination of reason to understand this background of disease cannot be attributed to a single causal condition. Because not only are these two bodies, the figures are in other organs including the brain. By 1880 the figures were:

Cerebral meningitis 265
Cerebral congestion 160
Cerebral hemorrhage 103

It is noteworthy that organic heart diseases reached 319 in a year in Mexico City. Enteritis and enterocolitis arrived in 1541. Pneumonia reached figure of 1197 in the year 1880. Statistics consulted until the year 1905 did not show significant changes, however there are other possible sources such as the flu, syphilis, cancers, the numbers of simple meningitis, brain congestion and cerebral hemorrhages increase in 1225 to 1120, for the years 1901, 1905, 1906 and 1912 you can see the emergence of a new nomenclature for diseases that are the scourge of the population. The decline in the year 1912 may be related to the effects of the 1910 revolution and its high rate of deaths around the country, although is not a direct expression is part of a social condition for understanding disease after other explanations where was passing the armed contingent, which may be the sexual order, however, these tables show the presence of diseases that are expressing a greater compilation for the system of life that the human body has[75].

Statistics show that there is a relationship with sadness, melancholy, loss of joy, emotionally is a society that lives with anger, melancholy,

---

[75]   Robert Mcca, "El poblamiento de México: de sus orígenes a la Revolución", In José Gómez de León Cruces Cecilia Rabel Romero (coordinators) *La población de México. Tendencias y perspectivas sociodemográficas hacia el siglo XXI*, FCE, CONAPO, México, 2001. pp.033-77.

sadness, and has in their midst the presence of unhealthy, the dwelling has no adequate ventilation, there is no health in waste handling, water pollution exists, air, and organic waste handling has not drained, the car collector, the drain down the street and the habit of throwing urine at the street are not a good start to the health. We can say that it is forming a reading of building the body of citizens where emotions are combined with a dualism as a philosophical principle and explanations are seen with the eye of positivism. However, the vast majority of people do not have access to this new logic, and the only resource is to not go beyond the immediacy of everyday life. Food is a good indicator of how relates the flavor with body, emotion and response, which builds the body. Thus, even though it agrees and regulations subjects the body process living by the subjects is not reached by this reading of wanting to establish in the process of life the principle of health, said the *bulletin board of health of Mexico City.*

> In the present state of science, there are two legislative measures that could be considered as bases for the prophylaxis of infectious diseases and infectious and contagious: isolation and disinfection required. The first kidnapping the individual patient contact from others and the second, destroying the infectious at times and other purifying rooms, clothes and other items that may have been contaminated, both measures practiced the most rigorous possible and in a consistent manner contribute to prevent the communication of these conditions from one to another people and to decrease therefore frequency[76].

One reason to understand statistics is just this argument. Provided that can be correlated with this factor, which means that today there is no way to refute this explanation. The poverty formula plus disease gives death, is confirmed. While this reading is good according to the logic that supports it. It is also true for what I intend to do, is not enough to say that poverty, bacteria correlated unbeatable, because the social process becomes

---

[76] *Boletín del consejo de salubridad del Distrito Federal*, Tomo II, septiembre 30 de 1882, números 3y 4 México, p. 44.

more complex in its correlation of interpretation. Deaths related to the brain, lung, intestine, and liver, if we can find a distinct correlation most complex processes in the Mexicans body construction of the XIX century and their correlation with the XX century this means opening explanatory spectrum of a conception of the body that exceeds the relationship of organs, bacteria and viruses, and forces us to search the background of an epistemology that does not include subjectivity, emotion, we can not find other readings in the process of health.

The statistics of death is the completion of a background process that has a more complex long-range trajectory, this means we can see how the process is building a social memory, made of a competitive society, establishing the market consumption and it can be correlated with the principle of food, housing, entertainment, leisure habits, the need to articulate this principle we demand a careful reading of data. We can interpret as a body is dying for a bacterial or infectious disease and find the cause of a very simple way. But we refuse to do a reading and, I believe that there is a social process trigger more complex than a virus or bacteria. In case of data of hepatitis and liver cirrhosis, the explanation is extended to the consumption of alcohol and is easy to give a way out of this situation by the disease and when we make the correlation with the discussions and agreements regulating the use, consumption and production of this by the authorities does not matter if Berlin or Paris, the human condition is the same, but their construction process is not the same and that is where we can establish them a clear distinction of how a culture is composed and assumes in and of subjects who have the collective history that links them in the process of being part of a community, of a class, a sector, the uniqueness of each social subject shows that there is a long process in each individual and their correlation with which it was formed and built organic, emotional and emotionally. This construction has some stories that allow us to make sense of this reality expressed in statistics. We can see that certain styles and ways of living are in reality normal subjects, this is up as normal, without being able to detect that the artificiality is a principle that creates compliance after generation, hence it is expected life and death were the ancestors as grandparents and parents. Cirrhosis is the case.

If by the year 1897 most dominant diseases were the nervous system and sense organs 103, the circulatory system 49, respiratory tract 229 and digestive 534 are the diseases that cause mortality. Statistics from previous years keep a relationship with a process of infectious diseases and nutritional especially epidemics, however you need to do a joint with the emotional order process to consider the correlation with the organs and their destruction, is that a body with emotion relationship can explain the complexity of a building time where emotion plays an important role in the direction in which to settle in the organ. The emotional state is related to depression, sadness, melancholy, seems to be the most dominant at the time, Porfirian society was in a phase or process of modernity, where uncertainty and rumors are an important in shaping life expectancy of people, this means that the respiratory; digestive are the places where they can specify the route of construction is something that can be correlated with the deaths and treatments. Without doubt it is a society under construction, the vision of science is a positivism that allows other options but not actions. So we can find writings that talk about health alternatives, which shows little certainty about the methods, even though by that time the vaccines are 100 years old. The emotional process cannot be interpreted outside the status of a historical time raising the possibility of a new form of response of the body. Let us look the other side of the body that is sick in the eyes of doctors. It cannot be reached the meeting point between body and emotion and that there is not. His logic of a physiology, a symptom which is moved by a virus or bacteria, it seems that does not allow the possibility of an item or other explanation, in that regard is consistent and well applied positive science, but in the eyes of the distance, one can see a relationship with the emotional process, that means they were present. It is only necessary to seek the path of articulation with the social, and family and the individual, the weight of the social in the years 1897, plays an important role to make sense of an emotional state of uncertainty, fear, vulnerability as embodied in the body. It is noteworthy that the bulletins of the Board of Health reported the alteration of food, beverages, coffee, milk, and butter. It's a dirty in these terms, a society or rather a depressed city with diseased lung, intestine and parallel diseases which has to do with a depression or low immune system, the condition of the relationship between city- body confronts us with the concept of macrobody, which combines macrocosm because a process that should be read with new eyes. Emotions and mortality rates have a deep historical relationship that is established as a normal transgenerational process in the body of individuals.

From the emotions or illness, in this logic we try is, like, no distinction is a principle that concrete construction process that we can track over time, that means that an emotion can ride the historic process of society, a family, as having a memory and a body that keep this process, it is possible to create spontaneous, is a construction that can be found in the process of a society. The impact of the actions that governments do have their effects on the daily lives of individuals on how accommodating is the expression of a construct of choice according to his personal family history. Thus, an emotional condition is expressed in the body because it is part of the body, cannot be invented; it is just a drive that cannot be neglected in the present.

The report of the *Boletín del consejero superior de la salubridad in 1897* in its section "remarks[77]" makes a description of the higher rate of disease corresponding to respiratory and digestive problems as predominant. Also noteworthy that 100 abortions were reported in January that year, a larger number in 22 to February and increased in 43 compared with the monthly period for abortions, in 8 years is 57, Clearly not say under what conditions made or reported abortions, they may be due to malnutrition, which leads to weakness of the uterus, a process of depression, low immune system, this means that the process of being depressed will not is a transient state, because when it has changed the principle of life is possible to understand how a body came to the defenseless and only preserves but does not give life. It is a bodily process that is linked to the self-regulating mechanism of the body of the woman whose emotional state that is the principle of self-regulating metabolic regulation. Is the construction of a response that allows the top to make sense of life under threat of death. Abortion is a response from a body that is weak, where there is no certain future and critical to the child,

---

[77] "Observaciones" In *Boletín del consejo superior de Salubridad.* 3 época, Tomo III julio 31 de 1897, núm. I, México, pp. 285-266.

building a deep contradiction with the principle of life. If this woman's body found in the principle of defending life by abortion means that society in which we live is dangerous for the existence, statistics show that rates increase, the margin may be related to other more complex processes as malnutrition and the process of existential crisis that exists at the time, uncertainty, ignorance, unhealthiness show that the principle of life is in danger of death. Women shall, in the space where her body lays more accurately a social condition that threatens the body manager. It may seem foolish to link with the emotional process, but that truth is a condition of establishing body by way of government policy. The construction that does a body at a time, in a space that is marked, influenced by the complex process of human relationships, where not only is subjectivity, emotions, nutrition, also as an individual constructs his election to a life situation that relates to the society in which they live. Perhaps it is worth tracing the process of construction of abortion and women's emotional state and how it links with the years and its increase, but we can also construct an interpretation from the partner, and the process of being a woman in those years, that reading should appear in the emotional process that can make sense of this death and how the individual lives, in this case, the woman in the years 1887-1897 and the figures continue to show that something happened, and no way to explain it by positivists. In 1898 the death rates were 252 acute bronchitis, bronchopneumonia 82 and pneumonia 218, and, of course, diarrhea and enteritis 228, showed that something in the body of Mexicans going on in the emotional field and last but not least the civil registry 100 abortions reported 10 stillbirths were recorded, was the month of April 1898. This is an interesting fact. 10 stillbirths, the question is why. Today we understand that a child dies in uterus because the mother is sad, depressed, afraid, malnutrition, abuse, or the mother's body faces a complex socially process, even attending a wake. The emotional reality to be built in those years is only the beginning of modernity and the construction of a railway system, the arrival of the textile industry or a complex social system where competition is not easy to compete from ignorance must remember that 87% of the population is illiterate and that is the element that cultivates fear of living, rumors become truth, misinformation is another element that enhances, anxiety, guilt, the idea of the sinfulness and that may threaten the life to come, that is the message that the mother sends to the baby and is not optimism, it is an uncertain future complex for bodies taht can not be built in relatively short answers. Witnessing a new body construction from the womb, which excited will lead to the future citizen who is able to survive three years. The statistical returns us to the social reality. Again we find that it is children with the highest number of mortality. The social condition prevails and destroys the body that can not regulate itself, to survive in Porfirian society that boasts of being modern, is to require adapted bodies, the bodies of the children give the answer or not to build new processes of adaptation. Dying little not only impacts the future life, also shows that building a life in these times of increased competitiveness only expresses the complexity of a network where emotions are to fulfill their missions make sense of a body that can not be explained with logic positivist. Undoubtedly, the presence of bacteria and viruses are a dimension that contributes to premature death, we should not ignore the fact that if a city depressed, unhealthy, its inhabitants will not have another mirror or a reflection of something better that introjection natural reality and just shows us that life is expected bleak, the city, the body, son, life, death seems to be doing something hard to see when the process is monocausal, this means that we need other reading angles.

So, women, men, children and elderly are discussed in the process of making lasting life certainly no reason to look for other process of health and mortality of the body of Mexicans. Crowding, poor health, housing also have this effect on death rates, but their relationship is with a body and emotion, lung-sadness, that means there is a complex process with the individual, Domingo, Orvañanos in 1898 said on housing and air quality:

> For human agglomeration, designate a meeting of human beings higher than it should have in its premises, given the capacity of that local. Also, the agglomeration of many houses in a relatively short space, it is considered detrimental to health, it brings all the dangers of human agglomerations, in such, we will deal with both the agglomeration of individuals in the parts of the room, as from the agglomeration of the rooms themselves.

The air of our rooms is vitiated by various causes, but mainly by our breath and products arising from the various substances used for lighting and fuel the air we breathe contains some amount of odorous substance, putrescent and toxic; also is charged, of moisture and contains five percent less oxygen than the air we breathe pure-five percent of carbonic acid in the propensity that occurs even in the largest crowds, is not what has importance to living beings, but the accumulation of the organic matter and excess moisture are both so heavy that make the atmosphere of a confined space and increased carbonic acid, which is relatively lightweight[78].

This element plays an important process and of course the organs such as lungs inhale oxygen not only in small proportion but they face a more complex route, as is the lack of oxygen in the body. The same Orvañanos recognized that:

If you go to one of those rooms late at night, almost suffocating a little while to stay there, the picture presented is truly horrible and disgusting, in a small room, lying, piled the parent, children of all ages and sexes, and many people outside the family. Here everything is done in the presence of all, not only modesty, until decency it is completely impossible, it seems like the atmosphere is sensual itself.

And continues:

Retention of individuals of our people in those places corrupted air, enervates their powers create in their body less resistance to morphic agents, with the result that including mortality reaches a truly remarkable figure[79].

---

[78] Domingo Orvañanos, "Algo sobre legislación sanitaria," In *Boletín del consejo superior de salubridad*. 3 época, Tomo III, mayo 31 de 1898, num11 México, pp. 298-299.
[79] Domingo Orvañanos, *op. Cit*. P. 301.

I did not have to say: the figures are a reality that is combined with a process called respiration and of course the body is the place where you get this route of oxygenation, which shows that the premature death of cells is a reality that conjugated with the emotional process that is linked to living spaces, it is fair to say that the poor are most affected by these circumstances, they live in spaces that do not even have windows. You can see that a body is built with the politics of modernity, overcrowding, low oxygen air, ignorance, malnutrition, and the exaltation of emotions that do not provide certainty, security, trust, fear, sadness, not are good combinations to see the process of building an individual's optimal body in the XIX century, either, child, woman, man or elder.

This means that the complexity of the social network that is constructed has a deep network where is necessary to make sense of an emotional process that can become a trigger for the new response of an organism that wants to adapt in time[80]. The lung and large intestine reflect more the process of building this body space where are showing that children from 0 to 5 years with the highest rate of mortality of 1894, (804), 1895 (628), 1896 (919) 1897 (573), 1898 (112). The figure was too high, without considering that the rates of abortions and stillbirths increased. Of course we can not speak of a single organ or emotion, the combination is more varied we find a relationship with the social, but then we have to build other views on this emotional relationship with a sense of what is sung in popular songs in the sayings and proverbs, phrases that can synthesize an emotional expression, this means, building roads made by subjects to take their emotionality or emotional conflict. Search builds a social outlet to the ravages puts us in the construction of processes that seem to have no link with the emotional or physical process of the subject. At that point we can speak of dimensionality that does not require one-dimensional narrow focus, that is to say emotional is not only relationship with the organ, but can be expressed in the daily activities of the subject, his self-observation may allow you to establish a knowledge of the process and build up new possibilities for reading your body or, to summarize the point of a phrase enclose his truth body that can and is compatible with others.

---

[80] José Olvera, "Aborto, nacidos, muertos" In *Boletín del consejo superior de salubridad*, 3 época, tomo V julio 31 de 1899, núm, 1, México, pp.8-13.

5

THE IDEA OF THE WORKING GROUP TO THE EXERCISE OF CIRCUMSTANCES (THE CASE OF PSYCHOLOGISTS OF FES-IZTCALA, UNAM)

INTRODUCTION.

Clarifying the concept that psychologists develop about their work, from his training and intervention with patients, involves an unclear problem. The current curriculum proposal that constitutes a professional identity is superseded by the new etiologies and epidemiologies. We are interested in the career of the graduates of the Escuela Nacional de Estudios Superiores (ENEP) Iztacala Universidad Nacional Autónoma de México (UNAM), today Facultad de Estudios Superiores (FES), Iztacala unit, between 1983 and 1990.

In the present study, we use the methodology of oral history, so psychologists were interviewed from graduates of the Psychology course, seeking the fountain of life, the actors and their process of transformation. Working with this concept of a living document, we forged an alliance with the process that constructs the subject to make changes in his life, especially the potential of why he took an election and no one in professional practice, allowing us to make a social fabric with the original approach and present.

Interviews were conducted with psychologists who were trained in the curriculum of the FES-Iztacala, to know their career after graduating, and if they have changed or kept the original conceptions.

Changes in design and professional practice are social and personal implications that require us to investigate the formation process that was launched in 1978 and its sclerosis at the reality of our time.

This search shows that professional practice is articulated with a working concept that cannot be separated from service and personal style. All respondents have had to modify the concepts that were formed in college, which meant a reappraisal of the social problem of the psychological. That is, the new psychological epidemiology cannot be explained by an absolute or totalizing reading, all indications are that the health of "mental" is a very complex field of study. Intervention strategies applied behavior analysis has not given answers to a population whose health demands are increasing. The articulation of the organic and emotional seems to be the solution for these psychologists. The break with a fragmented conception the body is an option to give them a new hope to patients with psychological disorders, which have increased in recent years. Faced with a sclerotic system, and not allow new interpretations or intervention strategies in solving and preventing problems, psychologists had to break away from their initial training, especially because it allowed them to see people and showed them only stimuli or responses. They found that's not how reality works, and that the complexity of human behavior cannot cover it with a theory. At this time, their tendency is to articulate the corporal problem with the emotional, and that opens new questions on vocational training and therapeutic practice of health workers. This trend, seen in our cultural context reveals that seeks to give and receive a service, which is psychologically important part of the body. The need to be heard, to have a partner, becomes an important principle to understand why "style" of conditioned responses cannot function in our culture.

I

The category "work" is read in accordance with the training you have, it may be from the political economy, anthropology, history, sociology, or from reflections of an epistemological, methodological, practical, also from field research, with manual and intellectual workers. The richness of the discussion continues today, and surely will not end, with the understanding that human activity is to be transforming and enriching the construction of new interpretations. But undoubtedly, the subject is the actor that allows us to make sense of this human activity that is not only the act of selling a commodity known as labor to extract surplus value. This idea of political economy has changed in our time. Technology has played an important role to stop studying the work as a process of historical order.

> The current crisis in turn, promotes a new kind of flexibility, mobility of labor by means of trends related to a *productive restructuring* that takes the basis of the robotization and, in particular, the introduction into the work processes of "flexible use equipment ". This is a new kind of accumulation that transforms modes of operation of capital in its early stages of production and circulation also entails the creation of new activities that tend to be productive in industrial forms and the transformation of "traditional" services[81].

Certainly not only brings new changes in the ways of extracting added value, also has an implication in lifestyles, especially in the body of workers, whether manual or intellectual, both organic and mental[82]. In a highly competitive society are manual workers damaged their body by a stereotyped style of work. The meeting with the clock, with the labor authorities and unplanned society, allow stress rates to be exacerbated, and thereby builds the possibility that the body develops several diseases, including gastritis, colitis, ulcers, hemorrhoids, scoliosis, implying a change in their sleep cycles, emotional states, depression, chronic fatigue, low

---

[81] Available at Mario Camarena Ocampo, *Jornaleros, tejedores y obreros. Historia social de los trabajadores textiles de San Ángel* (1850-1930), México, Plaza y Valdés, 2000, p.202 ; Warner Sombart, *El Burgués*, Madrid, Alianza Editorial, 1972,p.371; Gunter Wallraff, Cabeza de turco, Barcelona, Anagrama, 1994,p.238. Ricardo Cuéllar Romero, *De obrero "músculo" a obrero "intelectual" Modernización de la industria textil del algodón en México alrededor de los cincuenta*, México, CEAPAC ediciones, 2004,p.360.

[82] Cuéllar Romero, 2004,p.51.

sex drive ... and the list can be extended. In this scenario, it is unlikely to find that the work is the possibility of self-realization, the space where individuals can build option for their spiritual development. Life is going to travel to the workplace and on the job, it seems that the subject only thinks about work, has work dreams, invading his personal life, in a kind of alienation or disposal. Such work impairs the body quickly. The exposures to accidents and diet rich in animal protein and carbohydrates have to do their part to ensure that workers are becoming emaciated and potbellied. A lifestyle and only shows that the benefits of modernity have not been fair or just. The extraction of added value is the rule that you cannot break, no matter the cost. These and other effects are presented on intellectual workers, whom become ill from kidney, insomnia, depression, anxiety, migraine, become addicted to any derivative of caffeine, nicotine, white sugar, alcohol, etc. Not to mention that their bodies begin to deform, cholesterol accumulates and develop psychosomatic illnesses and chronic degenerative diseases. The effects of work in the body depend on the specialization, is any part that may deteriorate.

In the case of health workers, particularly psychologists need to go to their formation process and find a correlation with the circumstances of their professional practice. It is instructive to see them trained in a concept that tries to be collective or group in a competitive society and demanded by the cult of individualism. We believe that professional psychology was faced with the condition of its time, and that gave it a new perspective on its professional practice, which relied on a program instituted in the years 1975-1980 at the UNAM, specifically in the ENEP - Iztacala, FES-Iztacala today. The proposal had a significant absence: did not consider the history of psychology or the history of science in Mexico[83]

---

[83] That means declassification in the present, and that makes vulnerable any curriculum, in terms of being overtaken by reality, that happened to the curriculum of which I speak. Available. Sergio López Ramos (coord.), *Historia de la Psicología en México*. México, Editorial CEAPAC, 1995. This text is where we analyze the importance of the history of psychology for psychologists and a tour from 1840 to 1985 in Mexico. Armand Mattelart y Érik Neveu, *Introducción a los estudios culturales*, México, Paidós, 2004, p.175. This is a text that summarizes the studies on culture and the object of study in the field of Anthropology and it shows the process of diversity of

and that made their approaches are exceeded by reality in the midst of social and political crisis.

Make a curriculum oriented behavioral technology or technology education in a non-believer country of only symbols, but still believes that the blame for what happens is not related to his personal choice, but blames others for what happens, we should put on alert before the behaviorists attempt to innovate or accommodate reality to elicit a response to a stimulus. In this logic, the professional profile of a psychologist should be aimed at an exercise mastered the technique and provide solutions to the demands of the population. This requires us to go in search of psychologists to find out what happens to your practice at a distance of ten to twenty years of having graduated from college. Do they really use their theoretical and methodological tools to do what was promised, or there was another process that had to face to provide solutions to social demands and individual patients?

II

Make scientific psychology is only a proposal that has a look of such ideology. We cannot speak of a hegemonic proposal. This idea of the absolute in Psychology presents some problems, and one of them is to conduct a study and diagnosis for all individuals, rather illusory thing. Consider the complexity of cultural, religious and political of our time[84], therefore, define a social reality from behavioral objects as in 1980 by Emilio Ribes, reflects that he had come to despair at the necessity of beginning: self-affirmation prediction. It was necessary to train psychologists with a proposal that would, step by step, a reliable answer in the professional work.

You could not ask for more, a psychologist operationally planned training of the psychologist:

---

readings about social reality, which allows understanding that there is only one.

[84] Dejours, Christophe, *Trabajo y desgaste mental. Una contribución a la psicopatología del trabajo*, Buenos Aires, Ed. Humanitas 1990, p.241.

It's time to put aside our preconceptions about what is Psychology and address, with the rigor of laboratory research, evaluation and determination of the objectives we seek to establish training programs in psychology.

The answer or answers that we get for this is the product of a long process of research and analysis, hence the importance of discussion and you start planning early[85].

Bet on technology education, applied psychology in the eighties, materialize the dream of building a professional with an innovative curriculum, a window for psychologists who claimed the new Quijotes of children with impaired psychomotor development in Mexico. Psychology, therefore, be reduced to applied behavior analysis and that marked a conception of what is expected of graduates and teachers. Ribes said on the program:

> a) Links with vocational education priority social problems: b) Breaks with the schooling and elitist conception of the profession, the primary objective in planning the training of paraprofessionals and nonprofessionals; c) Defines behaviorally teaching situations, ensuring ownership of the learning tasks and curriculum content; d) Converts the evolution integral part of the teaching-learning: e) Specifies objectively necessary correspondence between the behaviors of teacher and student behaviors; f) Reorganizes the new way the traditional problems of psychology and provides a parametric conception of behavior, and g) Requires the effective participation of teachers and students[86].

---

[85] Ribes, Iñesta, Emilio, "La formación de profesionales e investigadores en psicología con base en objetivos definidos conductualmente", "El diseño curricular en la enseñanza superior desde una perspectiva conductual: historia de un caso", "Innovación educativa en enseñanza superior, reflexiones sobre una experiencia trunca", En Javier Urbina Soria (Comp.), *El psicólogo, formación, ejercicio profesional, prospectiva*, México, UNAM, 1989, p.338.

[86] Ribes, 1989, p. 359.

With this technology, is expected to give results in the field of deprofessionalization. Exercise Psychology faced a challenge: service. Towards the years 1980 -1985, social movements have shown that Mexican social reality could not be represented by a stimulus, nor was it possible to condition the consciousness, the reality did away with any theory derived from educational technology and the field of Psychology was no exception. If deprofessionalization began to gain ground, while it lost both: it was not possible to train functionally illiterate[87]. The project of building a shared knowledge with paraprofessionals was unsuccessful.

It was believed that users could be trained to apply the techniques of conditioning, it bypassed the cultural process, individuality, and, especially, was the underlying reality was one and the principles of behavior are universal: even the idea of collective work is a romantic assertion has no basis in the education of psychologists. The argument proposed by Ribes was free of odors in the social history of Mexican society:

Psychology faces in this context, two additional problems of its effectiveness in the social: 1) To capture the natural fields of application of its own as a scientific discipline and 2) To extend their professional competence in social, by fracture patterns of specialization that institutions have imposed[88].

For the author, the profession of psychologist and the profession in general, is the "monopoly of specialized intellectual work."

Undoubtedly, this training addresses the need to build an identity in the exercise and service proposal: for whom to work or who had to attend. Unaware that a human being cannot be matter of a definition by objectives lacked even think about what is meant by the psychological. The proposal to separate from the provisions of social institutions demand a different profile and a link to the service concept is different from what was believed to professional psychology. The demarcated areas were given by the ideals of a definition of objectives, a conception of remedial, without

---

[87] The Professional of Psychology, in that logic, not only had to share his knowledge with others, I must say it is a proposal that is similar to rehabilitation based on community, which is the logical ideas of cooperation, it does not talk about competition, attitude that cultivate the cosmopolitan societies.

[88] Ribes, 1989, p. 373.

the use of history to understand the present of the new professional. The imperative of objectivity rule out the explanation argued that the science of psychology should not only serve to make the human being to condition or acquiring new repertoires, it was necessary to go beyond the idea of conditioning and training.

The failure of this scheme, which did not include the historical process, is explained by their goals and impediments within the institutions, not allowed to make another type of intervention. If the model did not work had to find where was the fault, and so the problems listed Ribes in 1986, when the model had come into its worst crisis:

1. The professional demarcation institutionally imposed to psychology and the analysis and understanding of the social role it plays within the context of specialized intellectual work in different fields such as health, education, economics; etc. 2. Social epistemology or exterior of psychology and their interrelation with other sciences, disciplines and delineation of the natural fields of social application, based on scientific and methodological characterization. 3. The construction and evolution of a technology effectively applied, consistent with the theoretical model of basic scientific body and capable of being transferred and exercised by non-professionals who are generically called "community" - and 4. The formulation of a theoretical framework that allows identification of problems and tasks for which the technology must be designed, and its "fit" to the criteria and requirements prescribed by the very non-professional[89].

Make the design of a psychological intervention, where you think the transfer of knowledge to a community group can make sense of a social and political commitment. But if the technique is the means by which the objectives of a professional occurrence, well worth going to this definition of the technocrat:

(...) Which represents, therefore, instrumentalized reason that the concepts, emptied of their contents, have become mere shells, formal. This also explains its professional indifference in relation to the ideological purposes of the state or body, which constitute one of the gears. This does not mean at all that is depoliticized, but simply that the specificity of its role as waiver of any ideological position.

---

[89] Ribes,1989,p.374.

It can be used by a political system as the other, since it has to be structurally efficient, useful, and productive and nothing else. Their moods, the dimensions of psycho-cultural or political personality have no great importance[90].

Said W. says: "The characteristic of the modernist form is the strange juxtaposition of comedy and tragedy, of high and low, the current and the exotic, familiar and strange[91]." So I think it is this concept of the modern in a developing country. That is, the culture and historical processes are not of great relevance in the curriculum, science is weighted as the only resource for planning. The problems arising in practice are unpredictable, which means that a professional with the technocratic idea is to face the crisis that stems from the confrontation with the social and personal service users. This condition reveals the superficiality of a technique you want to impersonate a culture.

Since the technocrat, as the intellectual and religious, is a victim of "plated". Try plated models of development and productivity; unfortunately, hardly stick to the status of their environment. His love of numbers, statistics, quantitative criteria, their specific actions make him a being apart, real and unreal at the same time. Real because there are many technical achievements whose relationship is attributable him, but unreal because you have the impression that all these works raised to hurry, as under the influence of a magic wand, often impressive things pass by only rubbing them, giving the illusion of appearing to not scratch if you want the personality of the country, do not alter the constitution even profound of things[92].

The functional is embedded in the consciousness of what should be important, but the technique is only a means and logic of the planners, the figures are the only indicator of the success of his intervention. It follows from the need for correspondence between what psychologists have learned and what they do today in practice. Did they succeed with deprofessionalization? Do they solve the problems of the new social

---

[90]   Shayegan, Daryush, *La mirada Mutilada. Esquizofrenia cultural: Países tradicionales frente a la modernidad,* Barcelona, Peninsula,1990,p.186.

[91]   Said, W. *Cultura e imperialismo,* Barcelona, Anagrama, 1993, p.298.

[92]   Shayegan, 1990, p.189.

reality? What are the processes that have experienced professional to meet the new demands of the service? Perhaps the formation of a curriculum heavy on technique allowed them to meet the other in daily practice. How does a concept of work, you learn, with a viable technique is to provide solutions to the current epidemiological problem in the field of psychology? The numbers of suicide and children with learning disorders and marital problems and divorces have increased, the depression is expressed without range of age and chronic degenerative diseases (ulcers, colitis, asthma, diabetes, congestive renal) have become leading in national statistics. The concepts of the psychological work and establish forms of involvement and participation in finding solutions to the patient.

III

THE CONCEPT OF WORK

The practice of psychology in our time going through a crisis, because it works two key issues: the verbalization and conditioning. We can locate the therapies that work only with the word, which conceptualized the verbal and behavioral conditioning. They are held in a fragmentation, which deals with a body part regardless of everything.

This focus on the word or in the development of codes do not allow the complex world of human construction in geography, is a family space, and the process of appropriation by the individual meanings and representations that make his a microcosm of life, the body is the space where they express the social and cultural processes of their time. The new etiology of psychological problems cannot explain patterns of conditioning and psychoanalysis only; you must open the interpretive spectrum. It is not possible only to read micro part that is not related to the whole body and less to culture and social processes. The set of problems of modern societies relate to the body of individuals, this means, the body builds new answers to adapt to the abandonment of man-nature relationship. The human condition is artificial, and that prevents their adaptive responses to be fast. The breakdown of the cooperative network within the body is expressed in the forms of mismatch and lack of harmony with the geographical space. Hence, this new problem cannot be interpreted with a fragmentary scheme: this is the reality of our time and training of psychologists is to live this search phase. With the increase of disproportionate conflicts in psychological health, epidemiological studies show that going to the high demand for service. It is accepted that the vast majority of the problems of consultation in the Mexican Social Security Institute (IMSS) are of a chronic degenerative an emotional origin, hence the constant recurrence. The basket of drugs is not effective; it is known that the only thing it generates is iatrogenic. For psychology part of the same conception of the remedial, the results are meager.

In 1978, it was thought that professional psychology had to make a practice that you call deprofessionalization, and consisted on health professionals who had to participate in giving the knowledge to users to make use of the techniques they could lead to a cure.

By 1980, the Mexican social reality imposed on the ideals of the behavioral deprofessionalyzing of School of Advanced Studies Iztacala and, indeed, in 1981, when the economic crisis ended with the ideology of being in the horn of plenty for all people of Mexican society.

INTERVIEWED

The concept of work is derived from practice or work ethic, two levels that are not necessarily related, but articulated in a reading of what is done in practice. We find a discrepancy between what is expected from idealized or psychologist in social work and the reality he faced, his concept of work: the school told them to develop in various areas, of course, applying the psychological perspective, "the psychologist would help solve problems at different levels on the site is developed, scenarios such as school, clinic, rehabilitation centers, etc[93]". One thing is very repetitive in his speech:

> Had to work with things that if they looked and could be measured, that the matter of the soul, energy, spirit, and intangibles

---

[93]   Interview "Una promesa de trabajo", López, 2004. The informants interviewed for the development of this research will be kept under pseudonyms to protect their identity.

were not able to work with them scientifically and did not warrant any work; becoming then a behavior modifiers[94].

The definition of a line of thinking marks styles involvement and participation, and guidance on the degree of complexity of a psychological problem. Fix the emotional world can be confrontational when using a method that only bets on conditioning.

> The work was to find contingencial relationships and change these contingencies, establish programs, find answers and change stimuli and responses could be molded around and change everything, the world was plastic and clay and you could do whatever you wanted[95].

As if the subject had no history and could be changed by a molding:

> I brought with an idea of some behavioral problems, but divided, when I started working I face and realize that you can not respond to pieces because they arise a number of factors, problems there, but there was no way to reinstate and that made me start a search that begins with a concern on how to work because there was no answer on dealing with a population, that makes one start looking[96].

However, the reality is imposed on a logic that only seeks to adapt, and needs expressed in the principle of questioning what is done with a non-utilitarian sense of service and internal growth; despite being a logic of reproduction, the actions disrupt. "A reproductive activity that allows the development of man (...) is the response to a need of psychological order[97]." Broad and general concepts, complexity of the actions in the

---

94 Entrevista "Todo debe ser objetivo" López, 2004.
95 Entrevista "El mundo de plastilina" López, 2004.
96 Entrevista "De la necesidad a la creatividad" López, 2004.
97 Entrevista "Una promesa de trabajo", López, 2004.

service gives them another dimension "Being able to be me. The ability to find myself in the growth[98]." This means that the psychological work:

> Resignified and charged a dimension of life, for me the psychological goes beyond the behavioral cognitive and now life. I tried to work that way (behavioral) and always ends in a dead end, was very limited what could be done, even in the area I wanted to work, which was psychology applied to sports where the issue was to pull to pieces a movement worked, even in those places, I exhausted the possibilities[99].

The confrontation of the model with reality imposes a critical or distress to the helplessness of not knowing what to do with the other, especially when what was promised does not work one hundred percent and, worse, there is no maintenance of conditioning set on the subject.

> Work: is everything I do to be a better person, in every sense of my life, most important, is how I have to find things in my life, to bring new meaning to lives of people I work with, forms to find, help others[100].

The subject leads, then his eyes to the possibility of resignify in his career served, from the inconsistency in the scientific discourse and the reality of psychological work with the other: "Apply knowledge of the career, one comes away with the idea to apply in certain places, and when I go out I find that there is congruence between what is taught and reality, usually I am unarmed[101]" All that remains is the search for alternatives for the meaning of work.

The new epidemiology, including organic and emotional disorders requires a different concept of the body and its psychological processes:

---

[98] Entrevista "Todo debe ser objetivo" López, 2004.
[99] Entrevista "Todo debe ser objetivo" López, 2004.
[100] Entrevista "El mundo de plastilina", López, 2004.
[101] Entrevista "De la necesidad a la creatividad" López, 2004.

First I had to resize or give new meaning to psychology in a different way of how I had been taught in school (behavioral), the first was a theoretical level, academic, it paid off, but still left some missing. Searching psychology takes another dimension, because I had to learn through me, with a disease I have, that the psychological exceeded what I had said was not only the psychological behavior, is steeped in a process that is inscribed in people[102].

However, see the background of social problems does not preclude the individuality, which survives in the social and family life, so understanding the process of a culture cannot prevent from returning to the subject who is part of the immediate.

Yes, now I try to see relationships, people, I don't condition, what is rewarding is that sometimes people tell me they prefer to see me because they feel that I listen, if they do something for themselves, having a space unlike Social Security (IMSS) or the ISSSTE (Social Security Institute and Social Services for State Workers) or other institutions[103].

Sublimate practice, in the sense to deposit the condition of personal life, facilitates to maintain optimistic given the lack of a conception of psychology that does not include people, just conditioning and shaping, a discarnate technique that does not include the personal process of who applies it.

An opportunity to grow all the time, I say I am very lucky, I have the chance to be in the places where human condition collapses and I have two opportunities: to help and learn to avoid repeating those things in my family or myself. I love my job because I get to review myself and especially have contact with youth, it allows me to be talking and listening to myself at the same time[104].

---

[102] Entrevista "De la necesidad a la creatividad" López, 2004.
[103] Entrevista "Todo debe ser objetivo" López, 2004.
[104] Entrevista "El mundo de plastilina" López, 2004.

A closer expression to the idea of what it means practicing psychologist, can be found in the effects of an action based on the Cognitivism by one of the people interviewed.

> It complies with the concept of work as I conceive, I can develop the work activity as educational guidance, is the demand in the institution, it allows me to be in contact with students, parents and teachers, I can contribute so that they can solve problems which I am giving in the service, and that is positive and allows me to create job expectations, which are true and I can develop personally and socially, I like my job, and what pleases me most is that I can meet young people and I can lead them through this stage of his life, I see a lot of expectations of my work, is productive and will work differently and I'm interested in those people who are able to develop that capacity as people[105].

On the efficiency of a model, the fact is that the difference lies in the personal process. For some psychologists reject the implied conditioning model immersed in a constant crisis of dissatisfaction with their professional practice. Hence the search for other models that can integrate what they see in their daily work. It is interesting to see that the theory is to manage the five elements, which considers the human body has an analogy with the macrocosm, and was read as a microcosm that contains five elements: wood, fire, earth, metal, and water. At the same time have a relationship with the organs and emotions: liver, gallbladder with anger, heart and small intestine with joy, stomach, spleen and pancreas with anxiety, lung and large intestine with melancholy, and kidney and bladder with fear. This proposal leads to a more comprehensive building process that makes the individual.

> The results in young people are faster, in people with chronic conditions I could not remove them, but I surmised that people learn

---

[105] Entrevista "Una promesa de trabajo". López, 2004.

to listen, to me it is very important, as is heard and understood and has improved its quality of life[106].

Response options are not limited to the psychological, the same answer "diabetes, Parkinson's, young people with panic attacks, I make the diagnosis treatment (...) if people are so stuck in the emotions I give a massage to work that part[107]" The distinct areas of intervention does not justify the application of behavior analysis. This is the case of children with developmental problems:

> (...) I work in special education, I try to tell how this relationship will be structured beyond the genetic, biological, problems in the body, in the sense of family, emotionally, that make them stay in a state and make more complicated problems that could be solved more easily and find ways to unjam the knots in the relations between them[108].

Gone are the proposals to seek stimuli or responses, comprehensive reading of the subject includes a body which may well have "diabetes, gastritis, hyperactivity, respiratory problems, work with the body of children with seeds or pellets, therapy for parents, to resignify the retarded child, trying acupuncture for degenerative problems, cancer, AIDS, etcetera[109]."

Through such diverse problems, maintaining the correlation between the emotional and organic. The psychological is not limited to verbalization or shaping. The reading of the body is related to emotional processes, nutrition, and family. The complexity of the body is revealed to the terms of a lifestyle where social and family history emerges and is embodied in the individual, living document that cannot be read with the orthodoxy of positivism. "The job growth is an alternative therapy (refers to both the user and the one who serves his body, which may include a change

---

[106] Entrevista "Todo debe ser objetivo". López, 2004.
[107] Entrevista "Todo debe ser objetivo". López, 2004.
[108] Entrevista "Todo debe ser objetivo". López, 2004.
[109] Entrevista "El mundo de la plastilina". López, 2004.

in physical memory). The results are satisfactory in a maximum of ten to solve their problems for which they came[110]." It is understandable that the personal and social reality has connection points and is expressed in the inner growth of the therapist and user. This therapeutic work is not only a utilitarian sense, but it is articulated with an attitude that sees in the other the opportunity to build a subject who cannot mold or condition.

> Working in an educational institution, I can see the level of students, teachers and parents, which seeks to find situations that are affecting the problem and the ways they can be resolved, according to people involved weather is pupil, parent or teacher[111].

Human relation and diversity is not only the subjective construction, it involves a complex process that leads to the psychologists interviewed to see the reality beyond the technology: "At work I see how people learn to compete, fight with husband, wife, with the leaders, the children are very upset[112]". This reality includes the emotional, the effects on the body or deformation styles are made in living, which means taking reality as a showcase to allow us to influence and affect:

> (...) The work at school (FES-Iztacala) is disappointing, 80 percent of the population does not see the signs of psychological problems in the future, are prepared in mechanicalness, competition, do not realize the problems that are part of[113].

In this logic the future of psychology students is not encouraging, it means that the curriculum continues its initial proposal, though not the same, the variety of proposals has not been able to eliminate one hundred percent educational technology, of Hence the concern of those interviewed:

---

[110] Entrevista "De la necesidad a la creatividad". López, 2004.
[111] Entrevista "Una promesa de trabajo". López, 2004.
[112] Entrevista "Todo debe ser objetivo". López, 2004.
[113] Entrevista "De la necesidad a la creatividad". López, 2004.

(...) Students (Psychology) who are graduating and alumni, if not taken up this opportunity to train with other spaces, other logics, other ideas, do not have much future, it is a fact, many graduates came back saying that they are not hire and that there is no job and drive a (*pesera*) public transportation or things like that to survive, it seems to me that knowing what they know do not have many options for professional survival, and professional growth[114].

Thinking about the social as a space where there are no jobs, makes no good conjugation with the decline of a model that does not offer participation, and problems are part of a theoretical proposal that has come to chaos:

The new psychologists, very limited, very poor, because theoretically do not see all, they are in a plane picture, students want to work in industry, business in life is to be selfish, if the workers earn little, is because is their fault because they are so stupid, I tell them a psychologist talking like that and judging does not have much future[115].

The expression of a need not only is the reading of an emotional state, it is the combination of a proposal that seeks to give meaning to what is achieved by interpreting from other readings. For psychologists interviewed, the urge to read the psychological body from another angle is a reason for thought, it seems that the tunnel has no end:

They are at the beginning of a crisis because they are people who come back desperate at least to the therapy they manage, rational, spoken, have not solved problems like suicide, depression, get devalued, there is no future, I think the crisis is present, but not as vibrant in them, who do not make the connection, but the more immediate future to reality, will have to return to seek other options[116].

---

114 Entrevista "El mundo de plastilina". López, 2004.
115 Entrevista "Todo debe ser objetivo". López, 2004.
116 Entrevista "De la necesidad a la creatividad". López, 2004.

In the process of crisis there is no lack of optimist for psychologists: "I think they're good, can do things, I think all psychologists, when they leave school, but they are prepared but not enough to deal with many things[117]." Psychological work can be viewed from different angles, and its implications on the construction process in the individual and society, both are linked to the curriculum, and gives us a look that lets discuss what you want and what is done with professionals, the social impact of psychologists since 1980 and today has shown only that the problems for which they were formed have not met expectations in this world of new diseases and epidemiologies psychological.

CONCLUSION

Psychologists interviewed were formed on a theory that promised to make an accommodation with the other. Reality without considering the cultural and historical process, and his concept of work is based on conditionality, which implies the neglect of the other parts of the psychological problems. Changing the concept of intervention in the field of psychology involves having a different conception of the body and its relationship to the internal parts of the body, that is to say the construction of an internal harmony that includes emotional and relationship with the organs. This allows giving an explanation to the new etiologies and psychological pathologies in society today. With the understanding that the therapeutic use of verbalization and conditioning have come to a border that does not include new epidemiologies that have been built in a competitive society. The body is the space in which embodied the culture and emotions, and become the expression of a new physical reality. The study of the body with the theory of five elements can articulate the processes that are built in the organs and emotions.

This allows to give a new relationship to the psychological process of our time, you can find three types of relationships: The creative cycle is one of generation, The opposite force is the control cycle and Tonification cycle, where the Earth element is considered to be the fulcrum during periods of transition between organs and emotions, levels of involvement

---

[117] Entrevista "Una promesa de trabajo". López, 2004.

by the organ or emotion to the whole body only shows that it is not possible to continue the fragmentary explanations. What we can scale the psychological. The search for new explanations and interventions led to these psychologists to look into this explanatory scheme, including the emotional and organic, allowing them to talk about cooperation and work ethic, the recovery of the human. The current curriculum is not functional as thought in the eighties. One wonders if the process living graduates of ten or twenty years ago have some difference with the present. The answer is that there are some variants that are as cultural attaches, but not allowed to give clarity in the daily work. Problem solving is not a priority in this concept of service work. Thus, respondents sought explanations that articulate work, spirit of service, and defense of another work and psychological etiologies are resized, the epidemiology of a psychological nature that is taking on new dimensions.

Clarifying the concept of work led psychologists to consider that the resume of FES- Iztacala, UNAM, is different to the condition of its time, and its graduates have not found the link to their professional and social reality that are living: the new psychological reality can not be explained with diagrams of positivism, it is not possible to sustain the deprofessionalization and the team concept as a social service that approximates the social ministry of the Church, this means, a care work.

Finally, we cannot ignore the fact that individuals have the election process to the solutions of the problems of their profession or to be built in their time and space. But it takes a different look to give a new meaning to the reality that is built before your eyes; the epistemological break is a possibility, but the risks involve repeat, duplicate or just add concepts to say that it is working. However, the practical implications reveal another results, and that we can find on these psychologists who decide to break with the original scheme applied behavior analysis, and that approached them to a body with organs and emotions where you can find the relationship with space and time of the individual, even with their immediate geography, which opens a horizon in the exercise and reflection of psychological practice.

II

The problem

6

Two ways to see the body

In 1997 we published the book *Zen, acupuncture and psychology*. With it we answer some questions about the work of service to patients and share the experiences of a work that opens new questions about the body and brings us closer to the riddle of the psychosomatic. We live with and a psychology that deals with the problems of man wanting to find answers atomized in mind, in behavior in the unconscious, on interpersonal relationships. It has given unsatisfactory reading after learning the history and life of humans is constructed in a geographical area according to a time and it follows the peculiarity of the body and individuality. The proposal to make a generalization in humans becomes obsolete when building new forms of bodily feeling and view of existence: the symbols and meanings are lifestyles, body representations which deflect the spine, which is installed in an organ and make a life with psychosomatic problems. That's why health becomes a commodity, an instrument to establish markets and readings that are not of great depth in the slogan of live with dignity and it confronts us with the questions of the century that just ended. Industrialists and scientists promised us that in this century would bring the possibility for a brighter future, the century has ended and the promise was not fulfilled, there are only new forms of body building and that challenge existing diagnostic readings in allopathic medicine. The health workers sought only new drugs and sophisticated instruments to implement a remedial medicine, even explored the tree of life, but the surprise was great: did not find the answers, just flashes and shadows which suggest that the human being is not chromosome and the realization of a body. In both age-old questions of San Juan de la Cruz, Santa Teresa, Benito Espinoza, even the Bible in the Epistles of St. John and of course the teachings of Buddha, to name a few authors and texts that dug into the dark side of the human condition, speak of the soul, the passions, attachments and body memory. The rationalists did their job: they covered with reasoning and discourses that aspect of the body that is linked to the existence and the new way of feeling in our time, but despite that logical reasoning could not hide that there was something beyond the Cartesian machine, which has supported the weight of four hundred years.

Finally broke out the human side that has no place in reason, society, culture, came to explore new ways of feeling and the eve of new movement of the symbolic life that was close to the language of mystics like St. Juan de la Cruz or Espinoza: the word as a tool that reveals the inner attitude, which tells us that the body has a memory that does not share any interlocutor, even allows the work to hide, and when expressed do with movement or statism. Another option is to hide in the depths of the human condition and never show it in public; let him become the uncomfortable inhabitant of personal existence: grief, resentment, envy, a feeling of resentment or hatred for something that we hurt by the mere fact of not allowing men to be an expression of existence. The Catholic religion was a sophisticated body of repression and thus allowed the body to look for new ways of expression, that the brain works with information that was not expressed via verbal. Humans have many subterfuges to look the other way and not meet this condition for years and although this was accompanied by discomfort and no options there. The sentimental death and the death of the joy of life became a natural condition of human beings; suicide becomes the only way to escape a distress or frustration; it is a possibility to escape from the body; evasion on drugs, shelter in the physician, self-medication, physical neglect, exploit the body in sex, greed, ignorance, in the banality of life, only build a few possibilities to justify the actions of beings humans.

A little in depth and opening the range of the possibilities of our present, we take refuge in saying of the story that invites the resignation:

"Thus we lived." Just wait until it meets the following in the default attitude of life. It seems that there is no alternative: the sentence is served one day and then our body will be prey to the family and social circumstances.

This is constant and often reaffirmed by allopathic doctors for a lifestyle that appears to us as natural and normal; hence the deaths with much suffering, in pain, trying to justify an existence filled with no flavors. The rewards of life are not only an end in pain and terror because they will die, but also address the implications of a lifestyle as if it had only one possibility: there are no options, do not exist in everyday life. Further reading of a body to show us how it is blurred from birth to death: the fear of death becomes an enemy to be conquered by any way. The truth is there is no way to beat that little space of time we played as individuals on this planet. Therefore, our attitude on this matter may give us a quality of life without much suffering by the fact that we will die, until now the fountain of youth does not exist. The truth is that there are ways to die and you can choose to enter that world to overcome the ignorance that scares us and allows us to cling to small ideas about what life means.

The pain is entered in Catholic societies and does not see the reality of a world that is constantly changing, closing the possibilities of life without guilt; a body full of those ideas will not only make life much more painful, find new symptoms to build on feelings and resistance mechanisms. The depth of the beliefs can override the body's process and make life with a basic sense: suffering in the name of a concept.

All this comes about because over the years we have found these ideas are present in the vast majority of patients with whom we work; and the problem is not exclusive to one class, in particular, there is a culture that embraces the society of our time and with this we can see that it is just psychosomatic shelter of human beings before the immensity of messages they receive and do not know how to process, or those who can not even make an interpretation to much information reaching them. For us it was a critical process to discover that patients have little chance of leaving the body processes that have been built for years, that is, they need to work with, and in the body to make room for building body that allow him otherwise personal existence, that means working with emotions and feelings, with the desire for power and feeling to dominate and control others, as the body is also a place where noble sentiments are saved rather than as resentment and envy. That would always be a personal process within the structure of individuals ignored by us, but if we know its manifestations and that can be important to explore the path that ran down to the demonstration that is consuming the person: the psychosomatic.

This text is a reflection of a personal and social process of a body that develops according to the family and society where he lives; we can find the complexity that develops from infancy to old age. The interesting thing is the process that is lived and how we could change it or give hope to the lives of many people tormented by not knowing how to live in the present, or even for not having a life plan. The options do not exist in society today, as even health professionals have engaged in that part which is unhealthy marketing body. It seems that in all times have the same ideals prevailed on the cure and remedies for certain diseases covered with cure-all, as if that were possible today. That only shows that an effective medicine that can really offer a better quality of life is far from reality. More than four hundred years with a medical model that has appropriated the absolute truth about health. And some believe it is time to explore other options and to implement the idea of taking responsibility for our lives, since the state and the doctors have not covered this lawsuit. Just look at the indices of morality and make a correlation with the policies and pharmaceutical industries to explore the complicity of a market that does not lead to other options based on the population is illiterate and easily influenced, any campaign can hesitate to users, especially when the culture is available only to certain sectors of the population, because the information used to privilege deceives much of the population.

Now we try to break into the field of health questions that are still unanswered. For any healthcare worker is no secret that humans have entered a new phase in relation to their body and managing expectations about life, anxiety and stress in our time that is fair to say, were opened in the XIX century and was called "the century of emotional disorders," the XX century, no doubt, was the century of psychological problems. Many interpretations can be considered to give an explanation, there are from sociology, medicine, psychiatry, psychology. In short the entire Western thought is a worthy effort to bring a truth that does not defend life even though life is the reason to be. This is a paradox that is not answered in the objectivist thinking and everyone who makes a criticism may seem jealous, frustrated, or even enemies of life.

Without doubt, this leads us to determine what we mean by body, and if we accept that is more complex processes than physiological, that body is a complex network of cooperation and not moved by the simple principle of feeding and excrete to prevent poisoning, a very basic idea that many doctors grow. We believe that the horizon of bodily process we can not exclude the emotions and feelings as an important part in our time; and not only that, but the spiritual development of humans, combines in one body and symptoms and ways of getting sick, but there is not a specific methodology for a cure. However, there are new explorations of the body space that offers the possibility of building a new relationship between the body and individuality. I point this out because I think the dualistic division has stigmatized and if that is the feeling of people, so constructed and so live.

These reflections on the body and the importance of seeing with new eyes, what we are only confronts us with this discussion, on the bodily and psychosomatic, and leads us to the original idea of the epistemological problem, especially philosophical and practical.

We believe it is feasible bodywork from another logic than the Cartesian or Newtonian. Of course that demands great effort, so we made a detour to approach the body. First, accept that I could not apply the same diagnosis for all individuals, as each owns an individuality that demands to know his body building process. Second, it is necessary to make a diagnosis that not only weigh the psychological, but also includes emotional and affective process of life and the relationship that this process has with lifestyles; look for options on a methodology that is not anthropocentric.

Conceptualize human beings as part of a universe where their predecessors have done before has consequences in the present, is to find a relationship to a process that is not isolated, that is, we believe that the way we live, this entails a mark of that history of predation that is embodied in the human body.

Let me explain a little better: the destruction of natural resources is closely related to body condition; the body is a microcosm that is integrated into a harmonious relationship with water, wood, fire and metal. When these elements are destroyed on planet Earth, the body also experiences the consequences. The energy imbalance in people is the first of them, and therefore breaks or alters the relationship that the body has with its immediate surroundings, and their internal organs, resulting in alteration of the organs, and the predominance of some emotions that are constructed in relation to the circumstances of imbalance of the above items. It is not unknown that people who live in places where before there were trees, they may be more irritable or angry. Man lives in relation with his space and can even be born in places where the ecological balance has been broken and the body will develop new networks for adaptation; then emerge new personalities, and relationships with the body: is a complex process that epistemology, Cartesian can not account. Thus, we are witnessing a new physical reality that is linked to ecological degradation, loss of values and the inability of people to new responses in a deeply unequal society, full of symbols and meanings that can not interpret the right code to a culture with new representations and icons that represent different for all individuals.

Aware of this new social reality and virtual, we question the value of the instituted in the field of health and our first response is to share what we do with the other, give the experience to other who does not see what he does not want to see, explore with eagerness to learn, and build options to comply with a new body and emotional problems.

7

Psychosomatic diseases: an interpretation

I

What is it psychosomatic? The above is an obvious question for anyone who has approached today to discussions about health or has been a patient of the Mexican Social Security Institute (IMSS) or other similar dependence and the doctor on duty has told you the illness is psychosomatic, that installs you in the group who need a placebo treatment for which there is no cure. It is a stigma that can condemn insecurity or the conjecture of individual madness. You will often see from allopathic doctors when treatment fails, explanations classified into genetic disorders of nervous origin, menopausal or somatic. That is a response to conditions that do not respond to a basket of drugs in government institutions.

Times have changed and the human body is a space where new realities have materialized: the body dominated by emotions. One of the last definitions characterizes them thus: "psychosomatic diseases, strictly speaking, are those whose manifestation and development under demonstrable influence of psychosocial factors. This means that not everyone who suffers from asthma is ill psychosomatically, nor is it true that diabetics have a purely physical disease[118]."

---

[118] Boris Luban-Plozza, Waler Pöldinger, Friedebert Kröger, Kurt Laederach-Hofmann, *El enfermo psicosomático en la práctica*, Barcelona,

The psychosomatic ranking has become a form of labeling of Cartesian logic that cannot understand the complex process of individual construction. Exceeding the design of the machine-body analogy is only a possibility of reading to avoid falling into stigma and denial of new body realities in our time, where the world of emotions has gained a dominant space in the body:

"The psychosomatic disorders are different and can be classified into the following groups:

1. *Conversion symptoms.* The symptom acquires a symbolic character that is not understood by the patient. The manifestation of symptoms can be understood as an attempt to resolve the conflict. Conversion symptoms typically involve voluntary motor and sensory organs.
2. *Functional syndromes (somatoform).* Generally, it is impossible to detect changes in tissue. Contrary to what happens with the symptoms of conversion, in this case the particular symptom has no particular significance in the patient's life, but it is a nonspecific result of the disorder of bodily function, or perception.
3. Strictly psychosomatic illnesses (psychosomatosis). They are produced by a primary physical reaction against an experience conflict, accompanied by morphological changes of articulation and an organic disease. The preconditions (predisposition), an organic response *(a locus minoris resitentiae* should be understood within the individual history of the patient) can affect certain organs. From the point of view of medical history, this group includes seven classic pathological psychosomatic pictures: bronchial

---

Herder, 1995, p. 1. The relation with other disorders is more complex, especially when you know the effects of an emotional state of health and that escapes the classification schemes. "The sadness and anxiety can significantly alter the fit of sex hormones, and cause not only changes in sexual drive but also variations in the menstrual cycle. The duel, another state-dependent of a large brain processing, depresses the immune system, those affected are more prone to infection and - whether or not a direct consequence - to develop certain cancers. One can die from profound grief. Antonio Damisio, *El error de Descartes. La razón de las emociones*, México, Andrés Bello, 1996, p. 143.

asthma, ulcerative colitis, essential hypertension, hyperthyroidism, neurodermatitis, rheumatoid arthritis and duodenal ulcer[119]."

The manifestation and prevalence of emotions as new condition in individuals is built on a question that goes beyond the casual logic of immediate, formal thought, because the relationship with the organic does not allow the proposed remedial and least try the subject with the search for universal laws in the past and present:

"They continued considering the human being as a machine, that is inherently complex, but this way continued to profess a strong reductionism. Reductionism - the idea that all elementary electrochemical processes that occur inside - has dominated the thinking of more modern medical professionals, which have proclaimed nozzle their membership to psychosomatic school of thought[120]."

---

[119] Boris Luba-Plozza, Walter Pöldinger, Friedebert Kröger, Kurt Laederach-Hofmann, op. cit. "In addition to this classification of psychosomatic disorders there are others that are possible and common. The alternative is the most important attempt at classification by Engel (145) presented below:

• Psychogenic disorders (primary psychic phenomena without involving the body or Influenced only): conversion symptoms, hypochondriacal somatic reactions, reactions with psychopathological states.

• Psychophysiological disorders (somatic reactions for psychic influence in the broadest sense): concomitant physiological phenomena produced by emotions or mental states comparable; mental cause organic diseases.

• Strictly psychosomatic diseases (somaticpsychic-psychosomatic disorders), characterized by the following factors: first manifestation at any age (most numerous in adolescence), once triggered, development is chronic, simple or recurrent, psychological distress determining their appearance, usually specific psychodynamic conditions for the manifestation of concrete organic disease associated, the psychological characteristics of patients are notable for their consistency.

• Somatic disorders (psychological reactions against physical illness). Now on days also called disorders in the assimilation of the disease. You have to separate them, master the coping styles (styles for dealing with problems)."

[120] Larry Dossey, Tiempo, espacio y medicina, Prólogo de Fritjof Capra, México, Kairós, 1986, p. 35. "Any physicist, psychologist or sociologist who claimed to be scientific as naturally resort to the basic concepts of Newtonian physics, and many of them are still clinging to these concepts even now, when

II

The expression of emotions may be interpreted using the tools of causality; is necessary to resort to the logic of building a body as a microcosm where a reading can be done without the lens of causality, this is the border universal laws face new epistemological relationship to the ways of explaining the conditions and new symptomatology that construct subjects; options and ways to make life a new building process of corporal-social-spiritual are not clear for many individuals; culture and processes of symbolization meaning have given another turn to the body construction[121].

Living life has become ways to feel and see the new line incorporating lifestyles. The construction of new messages and images allow, when not prepared, a reading so chaotic and can not make an identification of what is desired bodily and feeds the lonely idea of a more complex reading world.

We can say that this is a new reality across individuals and thus puts the problem of the interpretation of new signs and symbols. The human body is not as free as positivists think, is subject to the culture and desires of others to build and continue the same interpretation of the world[122].

The psychosomatic becomes a new social reality to be assimilated in the population regardless of economic and cultural resources, is a new chance in life styles, levels of somatization become a way of life or

---

physicists have left far behind. "Larry Dossey, op. cit., p. 11.

[121] Some authors have explored some ways trying to combine or exceed the positivist concept of research; their work may be a possibility for rational thinking and achieve sensitivity on possible options within the body. You can consult Fritjof Schuon, De lo divino a lo humano, Barcelona, Sophia Perennis, 2000, 162 pp.; V.S. Ramachandran y Sandra Blakeslee, Fantasmas en el cerebro. Los misterios de la mente al descubierto, Barcelona, Debate Pensamiento, 351 pp.

[122] The possibility of escape from culture or history is impossible. There is a chance to transform the body and let the desires that built us to be the same or hegemonic. Historicizing this process leads us to the readings of the ways of feeling with the body. I developed the theme in Sergio López Ramos, Zen y cuerpo humano, México, CEAPAC- Verdehalago, 2000, 125 pp.

a pseudo new reality that individuals construct as a risk factor for ways to bring its existence, in spite authors like Hanna say: "The basic object of the somatic throughout our lives is to have more and more control on ourselves, learning to navigate the tensions and traumas of life as a cork rubbed on the waves[123]."

The language becomes a possibility to make individuals take refuge in words to rest the confusion of symptoms and build new responses in the body.

"Our language is psychosomatic. Almost all the phrases and words that express physical states are extracted from bodily experiences. The individual can understand what is graspable. This would give us a broad topic for dissertation that can be summarized as: the human being, for each experiment and each step of his conscience, is to use the way of the body. Being human is impossible to consciously take the principles that have fallen to the body. The body requires a tremendous relationship that usually causes us fear, but without this link we cannot contact the principle. This reasoning also leads to the recognition that you can not protect man from the disease[124]."

This implies an attitude to listen to the suffering; develops a new way of building the relationship with the body, which means that the ways to decipher the messages also contain a way to build the process of assimilation by the body. The problem becomes a concretion in the body; it means not being able to do other readings of the sensations, the body's messages are interpreted with the outline of a life that is controversial, difficult in these times of images and values that are articulated in different ways with the physical reality[125].

---

[123] Of course, he refers to his method, but the idea of control gets away just from this process of empowerment of individuality is an illusion that is beyond the control of established ways of doing life. His recommendation to go from "floating man" is just to avoid facing the election process. Hanna Thomas, Somática; *recuperar el control de la mente sobre el movimiento, la flexibilidad y la salud*, México, Yug, 1994, p. 31.

[124] Thorwald Dethlefsen y Rudiger Dalhke, *La enfermedad como camino*, México, Plaza y Janés. 1994, p.91.

[125] The cases of bulimia and obesity are ways to realize images and values in everyday life.

The multiplicity of forms and lifestyles in contemporary society is only an approach that puts individuals in the logic of not participating in the construction of society intentionally; depersonalization is established as a mechanism outside the individual, which loses the sense of his individuality. We are victims of circumstances and we have options for our lives; so we are assigned by the family and society what we do, is assumed to be true, it likely remains the subject of making his life and his personal style of living. The loss of the options not only in the social sense but in the process of individual choice becomes a phenomenon that does not allow ownership of the body, so that means a distance between the body and society, which is a contradiction to be this, the body, a socio-historical product.

The psychosomatic becomes a refuge from the conditions of existence that does not allow you to enjoy and engage life in this society, on this planet. Humans are the possibility with their bodies but the problem is not desire, but at the moment are constructed ways to make life a principle, which recovers the dignity and the possibility to live in dignity.

III

The somatic is a new cliché or a new shelter from the limitations of medical models to explain the process of the body from another logic, or break with the fragmentation and approach with more humanized individuals.

Here are some examples.

"Conscience, widely used and abused term in modern science, is alive, kicking. Long heritage of countless philosophers speculation about the relationship of body and mind, has become today a favorite of psychosomatic theorists who have sought to assert its own way the importance of mind over health. But psychosomatic theories have never finished work fine for the simple reason that their working model of how humans are designed simply not allowed to recognize the primacy of conscience. The psychosomatic concept never came to dispense with an objective accent, so it always remained linked, as the glove to the hand, the molecular theory of the origin of the disease. Psychosomatic theories never provided a satisfactory explanation of the influence of mind over the body, starting as started from a reductionist explanation of the mind[126]."

Any reading of the psychosomatic, this informal logic, can be found in speeches by authors such as Friedman[127], who seeks a relationship with psicoimmunology. He argues that the blood tissue, or blood, has become a second brain to build an immune system is autonomous and independent of the brain; to attack the body and not viruses or bacteria. The change in the genetic code of macrophages allows them to become aggressors of the body; meaning that the individual develops cancer or any other opportunistic infection. However, this explanatory proposal to seek the cause was not entirely true, that years later it is discovered that the brain in relation to emotional states is where this information is regulated and its relationship to physiological change articulates the emotional process of the subject and builds a distinct possibility of understanding the psychosomatic process; this means, searchs in the blood project and kidney filtration, combine the processes of life that can not be simplified to a point of physiology. We have, therefore, that the process of articulation between the emotional and body to the positivists is a very difficult puzzle to solve, for the simple reason that there is an epistemological reading of bodily; fragmentation as an explanatory device reaches its limit and not exceed the border of chance and that's just a principle that hinders other interpretations of fact.

However this may be a factor that can open new opportunities for questions to the positivists, because it goes beyond objectivity as a criterion. This reality for positivists is concretized in a disqualification of subjectivity; it seems that positivist's scientists found necessary to return to their statement, "is an epiphenomenon," and there will be until reality confronting them again.

Another important line of work was the search that began in United States in the years 1980-1990; it was called the decade of the brain. The possibilities of reading are varied, the history of physiology is an example

---

[126] Larry Dossey, Tiempos, espacio y medicina, Prólogo de Fritjof Capra, México, Kairós, 1986, p. 35.

[127] H. Wolf Fridman, El cerebro móvil. De la inmunidad al sistema inmune, México, FCE, 1991, 183 pp

of this; with the progress of the microscope, with the improvement of lenses and optical systems, anatomists were able to "differentiate cells of different sizes and shapes and thus study architecture own brain[128]". What contributed to the identification of new diagnostic systems for brain injury, was the use of laboratory techniques that allowed a positive diagnosis and topography to raise speculation about the idea of spirit possession, this is, given place to everyone: both exist.

"The classic diagnosis is based on the location of the lesion: to go from general to particular, the neurologist identifies the affected organs and systems from a number of signs or symptoms, coupled with the causes that produce them, and independent of them, are grouped into syndromes. While the neurologist from the last few years has new research facilities, his first step remains clinical. It must first examination of the different functions of the nervous system: sensory, motor, amnesic, and so on. After developing a positive and topographic diagnosis, spend an etiologic diagnosis. As relief in this way forward, using laboratory techniques that allow an objective type and location of lesions. Furthermore, currently available methods of functional exploration of the central nervous system that can provide valuable information to refine the diagnosis and evaluate treatment effects maintained[129]".

However, this technology was not the answer to psychosomatic problems, know the anatomy of the brain did not provide clarity about the phenomenon of building an organic condition and make it grow into perspective that dominates the microcosm called the body. This search returned more interpretations, "to explain the phenomena of recovery could be formulated as follows: part of the brain, which is not normally associated with a specific function, could be reprogrammed, or more

---

[128] Simón Brailowsky, G. Stein Donald y Bruno Will, El cerebro averiado. Plasticidad cerebral y recuperación funcional, México, Consejo Nacional de Ciencia y Tecnología, FCE, 1992, p. 29.

[129] Simón Brailowski, G. Stein Donald y Bruno Will, op. cit., p. 39. "Both modern medical imaging techniques as developed and applied to the animal take us, therefore, to conceive the brain as a dynamic entity that changes constantly." "Different fluctuations around average or to a mode; these also may prove more durable and correspond in this case, to phenomena of plasticity", p. 53.

precisely, to rephrase his genetic background to handle the functions of the injured area. In medical terms, this transfer function is dominated functional replacement[130]". It is worthwhile to say that in this field have opened numerous lines of research have borne some fruit and open fields to reflect on this monstrosity called brain. You can consult the work of Marilyn Ferguson on "The revolution of the brain" for a reading of the possibilities on the broad spectrum that begins with the revolution and the decade of the brain[131].

Another position that attempts to explain the ravages of the feelings and emotions in thought and body, is the holding by Antonio Damasio which states could not have developed without the biological regulation "It is even more surprising and novel that the absence of emotion and feeling is equally damaging, may compromise the rationality that makes us human instinct[132]". Subscribe to argue that the different layers of the brain, besides making the necessary functions for survival, also regulate the procedures of emotions. A chemical reaction is at the origin of a thought or a feeling that gives concreteness to human reason. Damasio goes a little beyond the union of the parts and find the correlation with the body:

"In the landscape of your body, objects are the viscera (heart, lungs, intestines, muscles), while the light and shadow, movement and sound,

---

[130] Simón Brailowsky, G. Stein Donald y Bruno Will, op. cit., p. 85. The authors add: "In human clinical practice, rarely, if ever, we are truly trying to repair nerve cells in the brain. The rehabilitation of brain injured patients is entrusted to a certain time or functional rehabilitation centers, and more specifically to medical care teams consist of re-educators, orthopedists, and sometimes to clinical psychologists, while within the medical profession, many people do not want to accept the idea that there is a possibility of brain nerve repair. Despite the almost proverbial pessimism by some doctors, we know cases of patients who have suffered a stroke or brain trauma that becomes invalid and recover its functions dramatically, sometimes after a long period, much longer in fact better than the majority of physicians are willing or able to continue the assessment of damages. "p. 83.

[131] Marilyn Ferguson, La revolución del cerebro, Madrid, Heptada, 1991, 433 pp.

[132] Antonio Damasio, El error de Descartes. La razón de las emociones, México, Andrés Bello, 1996, p. 12.

represent a point in the range of operation of these organs at any given time. Usually, a feeling is the 'view' that momentary body landscape. It has a specific content: the state of the body [133]"

Conception as a result of genetic inheritance and social interactions with the social environment, which gives a voluntary and involuntary charge to settle a body that can be represented in the brain via the construction that is built into this framework of neural processes and are in the mind and the body used as a standard[134]. There is no possibility of understanding the dissociated form; but is not a comprehensive reading of the body and thought process, the attempt to articulate the human brain and the rest of the body seems an unnecessary search because it coexists in a unit that has cooperated for thousands of years and gives the fruit this harmony of organization that produces ideas, excrement, history, feelings, diseases, among many other things. "I think, since it affects the brain, the body that provides more support and modulations: delivering a fundamental topic for brain representations[135]". Which means that may be affected by brain injury may not present the body by the loss of intellectual functions and verbal. It is necessary to build this vision the brain-body giving information on the involvement of a social norm that subjects them to a style of behavior, hence the possible injury may disrupt the social norm and loss of contact with the regulations and the proposed life; this feature is identifiable with people in a head injury, tumors, or neurological conditions and even the type of society we are born[136].

---

[133] Antonio Damasio, op. cit., p. 17.

[134] "Surprisingly, the mind exists in and for an integrated body: would not mind as if it were not for the interaction of body and brain during evolution, individual development and every moment of our lives. The mind had to refer first to the body; otherwise it could not have existed. "Antonio Damisio, op. cit., pp.16-17.

[135] Antonio Damisio, op.cit. p. 17.

[136] It persists in its articulation of the articulated: "Body and brain are inextricably integrated by biochemical and neural circuits that are targeted each other. Two main routes verify the interconnection. The first thing that comes to mind is made up of sensory and motor peripheral nerves, which transmit signals from all parts of the body to the brain and vice versa. The other route, more difficult to imagine even though is much older evolutionarily, is the blood: it

The correlation between disease, affects and emotions today is not discussed much, what matters is when and how you give the process of change in the body and in this case we are not talking about a somatization, speaking of the effect of a relationship of loss or emotional state, such as sadness, that change the biochemical metabolism of the body. This process is distinct from somatization, buildings and its manifestations are different also; the first effect is the gradual and sometimes fatal. Can be managed with therapy or support, usually people get over it, in the worst case die. The second involves a different process: the subject produces a symptom in an organ and becomes a chronic health problem that prevents him from taking the course of his life and is associated with some periods of crisis according to a problem that was not solve on his daily life[137]. Wanting to find the feelings in a chemical relationship may not be questionable, what matters is the reflection that it follows to understand the process of feeling in the human and especially the denial explains social processes and historical-cultural make organics can be transformed to another state in their chemical relation, that is, think about that possibility is constructed to have extended from the forms and mechanisms of the brain to build new relationships in the body; indeed, in this set of new forms and ways to build that occur with emotions and feelings in everyday life. Passing convention is something that is not in breach of rules or the questioning of established culture, it reached another level of expression that the individual manages to build his social environment and that leads him to develop other reading of what he sees and feels in symbolic representations. The unit is worn on the brain and body building is not fragmentation, the unit becomes a basic principle; the emotional and organic build together, we

---

carries chemical signals such as hormones, neurotransmitters and modulars. "Antonio Damasio, op. cit., p. 107.

[137] His reflection upon return is to consider that "a deep feeling is not what we feel when we jump of energy or when we are killed by a lost love, those are emotional body states. The corresponding deep feeling, however, the mood, "even though they relate it to the deep feeling, not exactly capture it. "When the deep feelings remain the same type for hours and days and not silently change the ebb and flow of the contents of thoughts, it is likely that the set of deep feelings contribute to a good mood, bad or indifferent". Antonio Damasio, op. cit., p. 175-176.

can not conceive of a body outside the scope of an emotion and vice versa. There is, I think, the possibility of building new ways of perception and ways of fruition. Of course, reading a chemical relationship has the burden of history and physiological relationships found with feelings and moods in close relationship with a substance produced by the body[138]. In this logic the subject is subjected to the basic processes of the brain and no correlation with social and cultural processes and even less with the individual processes, this means, the choice of individual lifestyles, and cannot be a way to interact with your body. We are, in any case, the victims of Dr. Brain.

In this idea can not escape conditionally to reflex responses and that is the negation of the possibilities of building on the subject, even does not allow the process of individual choice and that means only look in the brain and its relationship with the body. This body segregation or, rather, fragmentation, is not an option for finding new relationships of body with culture and society of its time, the atomitzation becomes a little richer choice of subjectivity and construction of new codes in the body.

IV

So far I have spoken of the psychosomatic process from Western medicine and I have presented some of my ideas, but my interest is in the theory of five elements posed by traditional Chinese medicine; holds that each element maintains a relationship with the organs, for example Kidney and bladder with water[139]. Having this in mind acupuncture works. I turn to it to fully understand the psychosomatic process.

---

[138] "Long time ago we know what chemicals can alter emotions and moods: alcohol, narcotics, and a host of pharmacologists agents can change the feelings. The known relationship between chemistry and feeling prepared to scientists and the public to the discovery that the body produces chemicals that cause a similar effect. Today it is accepted without difficulty that endorphins are the brain's own morphine, which can easily modify our feelings about ourselves, and the world of pain. Also acknowledges that the case with the effects of neurotransmitters such as dopamine, norepinephrine and serotonin and neuropeptide modulators. "Antonio Damasio, op. cit., p. 185-186.

[139] The five elements are wood, fire, earth, metal and water, which in turn

The conception of the body that manages the sustained acupuncture on analogy with a microcosm of harmony which, when violated, causes energy imbalances in the body, the imbalance can have diverse backgrounds ranging from food consumed, the way you breathe, the place where one lives, the type of work, ways of expressing feelings and sexuality as it is lived, among other things. Now just to give you some ideas. The analogy with a microcosm breaks with the idea atomized, anthropocentric, to untie the man of the place where he lives and make him think that he bilt its own "nature" and is independent of the processes of the planet. This illusion has led only to that individuals have a hard time finding a place on the planet, in society or something closer: their life. Failure to cultivate the relationship with nature has ecological and corporal imbalances that are expressed in a new complexity reason-body which process attempted to account for Western theories.

The whole discussion about the brain is the result of this disconnection and fragmentation with the man himself. So try to interpret the psychosomatic process in a concept of harmony as a microcosm of the human body. One of the major health problems in recent years is related to emotional processes and their impact on contemporary societies[140]. We seek to understand the process of the microcosm and the relationship established with a body and emotion: I will try to simplify for exhibition

---

are related in that order with each item: liver - gallbladder, heart - small intestine, spleen - pancreas, stomach, lung – large intestine, kidney -bladder. Turn them into three types of relationship: 1) generation, 2) control cycle and 3) tonification cycle. Each has a chance to life of individuals and their emotions. The reader may ask why not give priority to the brain or body. Simple reason that cannot fragment this little microcosm that answers, yet, to seasonal changes. Are available for extension of the theme: Ye Chenggu, *Tratamiento de las enfermedades mentales por acupuntura y moxibustión*, Madrid, Miraguano, 1991, 190 pp., y *Fundamentos de acupuntura y moxibustión* de China, Beijing, en Lenguas extranjeras, 1984, 461 pp.

[140] The following text is an example of this. Robert Desjarlais, León Eisenberg, Byron Good y Kleinman, *Salud mental en el mundo. Problemas y prioridades en poblaciones de bajos ingresos*, Estados Unidos, organización Panamericana de la Salud, Organización Mundial de la Salud, 1997, 590 pp.

purposes. Acupuncture holds that an emotion or feeling is related to an organ and vice versa. When the dominant emotion is installed in the body, which allows the individual developing energy disorders in him. It breaks the balance is kept between the body and emotion beyond that idea can extend the concept and understand that the organs are related, that is, they are not autonomous, which means that the energy imbalance [141].

In this logic anything that happens to a person is outside his condition. An example: the relationship of fear with the kidney is may be of two forms: one, that fear is rational, this means, in the head and kidneys can suffer its effects: swollen, is susceptible to infections, and two, that kidney had an infection, then the person has some unexplained fears, especially fears of death, loneliness and in relation to this change are articulated menstrual changes, problems of low libido and fatigue, in some chronic cases. The breakdown of energy balance is closely related to living conditions and the ways that express their feelings, how they live, with or without stress, body care, type of work and food, the ways in which are stored grudges, emotional abandonment, over-protection, resentments, etc., in family relationship. This is important because it depends on how the person develops his psychosomatic process: deposits it into an organ, joint or produce any symptom that may well be migraine, hemorrhoids, constipation, asthma, chronic depression or sadness, in some cases psoriasis problems, including less common. Well, when we face this symptom we search the relationship to the organ or organs in some cases, because the relationship between some cases is between mother-child, which means that if an organ is affected, this may turn affect other, which the microcosm disharmony. In that sense nothing happens to the body is different to the interpretation of relationships that allow finding the depth of organic and emotional problems.

---

[141] It is held in acupuncture that there are two types of energy: a genetic and acquired, the first is that we inherit from the parents and the second is acquired through food, air and liquids. Good management consists to eat foods that are nutritionally balanced and rational use of sexuality, to prevent the loss of K i genetic. Available at: Daniel Reid, *Los tres tesoros de la salud*, España, Urano, 1993, 507 pp.

So far we can see this process in our patients and the restoration of energy balance is the ability of health. This aspect of interpretation is a first step; the second is when doing the treatment and recommendations to restore energy balance, which can range from enrichment of the diet to be given a chance to listen; that will bring the possibility of better quality of life. I consider this aspect of interpretation is of vital importance for not looking this vicious circle of the brain and reason.

8

An approach to the epistemology of psychosomatic

Etymologically, the word *psychosomatic* is composed of two Greek words: *psyche*, "soul", and *somatic*, the "bodily" in an animated being. It involves a material, physical or chemical symptom, dependent of a solid or liquid alteration of the body to differentiate it from the functional system[142]. This means that both the soul and the corporal exist in one body but social and personal processes can alter their functionality. This unit has been read and interpreted from different epistemologies. The world's cultures are a living example of this: the body is experienced and read in relation to geography and the various rituals that are built around their secretions, their cycles, their decay, death, etc. For now we are interested in the reading process that makes medical science on this duality called psychosomatic, for two reasons: *1)* is a classification that is used as a resource for what, with respect to disease, can not explain the dominant hegemonic medical model. This makes two possible ways of evasion of the problem: one, throw the tray of genetics, and other say that the condition is genetic, so they refuse other interpretations or readings of the body with respect to their signs and symptoms. *2)* The psychosomatic processes have been reduced to root causes of social-cultural, biopsychosocial, chemical, emotional-chemical, blood and tissue expression of a low

---

[142] *Diccionario léxico hispano*, 2 t., México, W.M. Jackson, Inc., 1982.

immune system, and culminating in creating the new discipline called *psychoimmunology*, among others stem from a growing problem, such as psychosomatic illnesses.

The rates of diseases involving the emotional control anywhere on the body have increased over the past two decades. It is known that the problems of a psychological, psychiatric and psychosomatic order go to high for various social conditions, political-economic, war, chronic malnutrition, stress, crowding and overcrowding, loss of expectation of personal fulfillment, family crisis and values in contemporary societies, ecological degradation, among other possible reasons that give us a very gloomy assessment creating new diseases in individuals[143]. Perhaps an important aspect demanded by the problem is that the psychological, mental and psychosomatic diseases do not swell the death statistics to impress politicians or be an instrument that can be used to pressure responsible for health. Opposite is true: they are silent and their growth is becoming so common in the population, is common to accept that someone has gastritis, ulcers, hemorrhoids, migraine, constipation, tachycardia, insomnia or disorder treated with medication for years without obtaining a lasting result. Recidivism is a lifestyle, is incorporated into everyday life without much hindrance and individuals learn to live with certain conditions. Mental illness has been characterized as an area to which the

---

[143] The literature is very coarse, ranging from self-help books to exploring new alternatives in the treatment of diseases to reduce or attempt to cure their symptoms and suffering, and even works that address the social and family aspect. Ryosuke Uriu, *La medicina, agradable*, México, Herbal, 1997, 233 pp. Margarita Estrada Iguíniz (coord.), *Familias en la crisis*, México, CIESAS Antropológicas, 1999, 102 pp. Jean Didier Vincent, Biología de las pasiones, Barcelona, 1988, 331 pp. Hugo Liaño, *Cerebro de hombre, cerebro de mujer*, Barcelona, Grupo Zeta, 291 pp. Michel Bernard, *El cuerpo. Un fenómeno ambivalente*, Barcelona, Paidós, 228 pp. Raimon Panikkar, *El mundanal silencio*, Barcelona, Martínez Roca, 1999, 174 pp. Antonio Munné, *La evidencia del cuerpo. Como llegar al equilibrio cuerpo, mente, espíritu*, Barcelona, Paidós. 346 pp. Michael Samuels, *Creatividad curativa. Como desarrollar nuestra creatividad oculta para curarnos y curar a los otros*, Argentina, Vergara, 2000, 382 pp. Jon Elster, *Sobre las pasiones. Emociones, adicción y conducta humana*, Barcelona, Paidós transiciones, 208 pp.

health workers only understand as possible product of a psychological affliction that relate to unhappiness. Still in the seventies and even today[144], the etiology was sought and is sought in the biological and genetic aspects, leaving aside the social and cultural processes that dominate the global and national context[145]. The World Bank reports of 1993 said that depression was ranked fifth among women and seventh among men[146], probably today these figures have not decreased. It seems that the growth of these and other related health problems are increasing psychosomatic processes[147]. Assails me a question: What is being done? To answer I searched the medical and experimental literature. In 1968, Niel U. Miller and M. Banuazizi made some attempts to find relationships between organs and emotions; used the experimental scheme via the conditionality to restore intestinal contraction response in rats by brain stimulation. His first conclusion argued: "cooperative learning of visceral reactions can exist[148]". An interesting premise on learning about organs to regulate

---

[144] F. Bolívar Zapata, *La genética moderna: horizontes*, México, El colegio Nacional, 1995, 79 pp.

[145] The problem is not only of an individual. Implies that "the disease impairs your ability to function within the family, work and social life. The worst consequence of depression is suicide. For example, in the United States is estimated that between 40 and 70% of suicide victims suffered from severe depression. Making a chain of effects complicates the explanation of an event such as mental problems, the tool to explain this process is the Cartesian model, it has reached its boundary, hence the need to see with different eyes these processes of health. Desjarlais, Robert, Eisenberg, León, Good, Byron, Kleinman Arthur. *Salud mental en el mundo. Problemas y prioridades en población de bajos ingresos*, Washington, Organización Panamericana de la Salud, oficina regional de la Organización mundial de la Salud, 1997, p. 49.

[146] Ibid, p. 90.

[147] reports from health workers always have chronic patients who do not go out for consultation and sometimes are handled with placebo, confirming the dominance of the emotions in the body. In this regard you can consult: Paul, Martín, *Enfermar o curar por la mente. El cerebro y el sistema inmunitario*, España, Temas de Debate, 1997, 501 pp.

[148] Quoted in Tomio, Hirai, *La meditación Zen como terapia. Las evidencias científicas de los efectos del Zazen en la mente y en el cuerpo*, España, Ibis, 1994, p. 129.

homeostasis through stimulation of new internal network responses; Miller went further, "has emphasized that homeostatic regulation is not limited to innate mechanisms, but can be achieved through cooperative learning visceral reactions[149]." While it would have to make the learning process deviated from an optimal homeostatic cycle serve as motivation for cooperative learning and conversely, restore the visceral reactions learning to return to optimum level. The author continues: "We believe that an innate control of these characteristics must be demonstrated both cooperative learning and pathological conditions, which are final products of the interference in that control[150]." That is, they opened a door to reflect on the processes of health and disease as is possible to construct an optimum relationship within the body. Homeostasis and innate cycle is broken because you can think about establishing a cooperative learning with the brain and organs. In 1969, Elmer Green said that "the electroencephalography feedback can lead to altered states of consciousness, and if so, the mental state is very peculiar and goes beyond the normal thinking[151]". Dr. Hirai suggests that this is an attempt to view the body and mind as a unit. In 1971 Kimaya experiments "revealed that it is possible to learn to control the alpha waves by receiving continuous audible or visual stimuli, even without any specific instruction. People who come to master the art of voluntary control their brainwaves can control the abundance and frequency of alpha waves, the left and right domain of alpha waves in the right and left hemispheres. While there are individual differences, the fact is that subjects tend to associate changes in brain electrical activity with verbal descriptions of specific mental states. Words such as *calm, relaxed, alert and aware inside* seem to suggest the mental state associated with the continued abundance of alpha waves[152]". Thus, we find that feedback and voluntary control are attempts to articulate the unarticulated by a reading of the body that only try to view fragments. A vision that tries to go beyond the representation of data is to Drouot, who says:

---

[149] Quoted in Tomio, Hirai, op. cit., p. 130-131.

[150] Ibid., p. 131.

[151] Ibid., pp. 133

[152] Ibid., pp. 132

"Correct a disease is directly linked to body function, healing, again corresponds to a transformation of spirit. In general, we can say that any healing process involves three interrelated aspects:

- he person, his attitude and belief system.
- The therapist, a doctor, psychiatrist or not, their capabilities, their attitude and their own belief system.
- The potential treatment itself[153]". But the problem may be diluted if only we seek in these processes; I think it is necessary to go a little beyond the traditional medical concept against the search of angst before so many lives lost. In that vein we find systems that work to relieve stress or make life easier competitive. Khalsa found "that the physical elements of therapeutic meditation can alter the mind and emotions. Moreover, means that the elements can alter mental body[154]". Confirmation justified with playful use of Eastern techniques in the West.

These experiments are the answer to the slogan of knowing what happens to psychosomatic processes of individuals, that today but no merit adjustment, however, implies a different conception of therapy. Perhaps it is worth reflecting on cooperation with its immediate space without undermining their self-esteem, that is, they can maintain their identity, their idiosyncrasies. Given the circumstances of our time is becoming increasingly difficult for individuals to maintain their mental health, balance, harmony with their surroundings, and no indication that will be different in the years ahead.

The medical researchers argue "we speak of psychosomatic illness only when a psychosomatic disorder has led to organ failure and thereby psychosomatic illness. Try this connection is difficult and depends on the principle that diseases were classified clearly before psychosomatic, although studies in the field of promise psychoimmunology close this gap

---

[153] Patrick Drouot, Sanación espiritual e inmortalidad. Las vías terapéuticas para el futuro, Barcelona, Luciérnaga, 1994, p. 132.

[154] D. Singh, Khalsa y Cameron Statu, *La meditación como medicina*, Barcelona, Diagonal grup 62, 2001, p. 42.

in the near future[155]". I find motivating discussions on the origin of the psychosomatic; we can find statements about the origins of sociocultural, intersubjective relations, depression, fatigue, fear, a lifestyle with high stress, etc. While that may be initially difficult to accept the installation inside the body because the symptomology with no age, as we present a child who is an old man.

So far we can see that it is not clear where is the problem of the psychosomatic, are approximations which are held in an epistemology that has concepts of *body, nature and health* atomized, this means, man is removed from its fragmented nature and its social nature. This entry is evidence of joint problem with a complex process as the human condition. A human being that sees a body apart from his historical process cannot be in harmony in his daily existence. The naturalization of the human body is present in medicine and the psychologist, build relationships of conditionality in the organs through learning or trying to make a condition suited to the circumstances of the body is the denial of representation and meaning to allow building more complex processes in thinking and feeling individuals. With these Darwinian epistemology and Cartesian is conceivable to think that the response wants to accommodate what is untidy. An epistemology of psychosomatic seeking phylogeny and ontogeny in human presents a great risk of making many experiments to make self-affirmation predictions of the fact. The molding of the body and its responsiveness allow expression of the target and possible manipulation of the physiological processes involving the brain and organs, so it is not easy to refute hundreds of experiments where physiological processes demonstrate their ability to be irreversible. The body and its capacity to respond to any circumstance become a pseudo true that experimentalists used as absolute knowledge as the problem of subjectivity used to the discursive, constitute underpin relationships with research experience of laboratory, so I explained as psychosomatic a chemical or physiological process; in the best of that field is removed and thrown into the bag of the biopsychosocial where you can talk about a doctor-patient relationship or the instituted as problem to solve, in short, is as large or extensive speech

---

[155]   Boris Luban-Plozza, Kröger Walter, Laederach Friedebert y Kurt Hofmann, *El enfermo psicosomático en la práctica*, Barcelona, Herder, 1995, p. 3.

according to the fears and anxieties that causes ignorance of any topic[156]. The fragmentary interpretation of this phenomenon corresponds to the long heritage of reasoning that has its history since the Greeks. The word *psyche* today means psychological, and includes such mental problems and all kind of leads in the theories and interruptions on the plight of individuals in modern society. The fragmentation or division that modern science has established for the study of the body faces an overspecialization process, which justifies the emergence of new specialties with the same logic of causality; this means finding correlations in a process where the reading of corporal reality is exclusive, this means, these self-affirmation predictions of fact and epistemology are based on a human body that seems to have no links with social and historical processes, except with representation and meaning that individuals have. Far, then, is the intricacy of the choice of subject and how it constructs the symptoms in his body, so I consider necessary to do a reading of the psychosomatic with another epistemology of body[157]. It must be said that the discussion has been addressed with the idea of diversity, but the questioning of reality do not want our escape, demand considerations other than what was common in our cultures. "We are obsessed by time but time no longer belongs to our being, but it has almost been reduced to a commodity[158]". Knowing this we faced a diversity of problems that point to the human condition, so the discussion of alternatives was attenuated by a hegemony that resembles a pitcher knowing where everything can be accommodated (one can fit everything

---

[156] Paul Martín, *Enfermar o curar por la mente. El cerebro y el sistema inmunitario*, España, temas de Debate. 1997, 501 pp.

[157] Necessarily with a reading of social issues including the unit or loop. As Castoriadis states. Looking for the meaning, when this is lost, many things happen to the body in relation to the internal construction of the organs, loss of sense of cooperation is an imbalance between organs. "The link between the psychological and social root, {...} is the socialization process imposed on the psyche, through which the psyche is forced to accept society and reality, while sailing society, with its limitations, by necessity, primary of the psyche: the need for meaning. La necesidad de sentido". Cornelius Castoriadis, *Figuras de lo pensable*, Argentina, Fondo de Cultura Económica, 2001, p. 187.

[158] Raimon Panikkar, *El mundanal silencio*, Martínez Roca, 1999, p. 40.

in small places if items are in order). "Diversity was seen like a, what do I prevent to see to other epistemic worlds, cultural traditions and more; other worlds that affect the construction of the subject as a being who lives gender, sexuality, social and intimate attitudes, shared life systems coexistence, politics, ethnic groups and societies of various kinds [159]" Is just an epistemology that emerged as a single option to interpret the complexity of relationships that are being built in today's society where the weighting of anthropocentric can be criticized and the body remains the point of discussion, only now with the idea of diversity, respect and the derivation of fields should know that are explored to investigate what happens to the individual. The idea of studying mankind has culminated in the romantic banner of utopians who now hope to understand the relationship of emotions to live better[160].

So for that we must seek, if the problem is a major capital? One of the basic principles to attempt a different epistemological reading of the human body is his conception and start from the idea of not accepting that it is just a conglomerate of organs, bones, a physiological or a biochemical process, not history and emotions, much less religious beliefs or just what you eat.

I believe that man is a living being that is part of the planet and can be understood from the elements that make up life on earth, for that reason, I can say that the human body has within it a balance - allopathic medicine calls it *hemostatic cycle* - corresponding with the exterior, this means, the harmony of nature can exist in the body, the complication begins when the external harmony is broken by the irrational use of resources in the industrial society that exploited forests contaminated water, earth,

---

[159] Rafael Pérez-Taylor, (compilador), *Antropología y complejidad*, Barcelona, Gedisa, 2002, p.

[160] "Every feeling is, therefore, a change in the state of the subject and the body in general, after the relationship with the object that stimulates it: changes in the body result in alterations in respiration, heart rate, dryness or humidity of the mucous membranes, urinary retention, etc., Carlos Castilla del Pino et al., *El odio*, Barcelona, Tusquets, 2002, p. 13 "I hope that the better we know the way we operate our emotions can enrich our lives, rather than diminished them", Dylan Evans, Emoción. *La ciencia del sentimiento*, España, 2002, pp. 12-13.

metal overexploitation and overheats with fire or specific geographic radiation. These social and historical processes are involved not only in the cooperation networks of plants and other living things, their effects are also expressed in the human body that has to face the dilemma of adapting, having a chronic illness or die[161]. Construction of options within the body is an interesting process that explains the psychosomatic phenomenon. If we hold that the body is a microcosm means that inside there is nothing without a relationship with the outside, there is an internal communication can not be avoided for the sake of objectivism, nothing that happens outside is foreign to the body. If we delve a bit into this relationship we find that the body is a space where not only brings together the elements of the planet, there is also a relationship within which individuals can build a time in the articulated body to the process of organs, which have a close relationship with emotions such as anger, joy, sadness, fear and intelligence.

A more complex process of this relationship is when crossed with the culture of a specific geography, that builds bodies with different internal networks of cooperation and competition, resulting in degenerative processes or the maintenance of health in an equilibrium process in the organism. The combination of the above with the family and feeding systems deed or complements a microcosm of tastes, smells, feelings, beliefs, flavors, ways of expressing emotions, socialization, and in particular styles of handling of emotions, which allows reading of a body from the ways in which the body think adaptive responses, this fmeans, lifestyle, individuality and expression of a way of being.

In this relationship, the organs with the emotions is where we can find the construction of a psychosomatic process to be interpreted as expressing

---

[161] The concept of microcosm we use in this work is far from the authors but converge on the idea of the relationship with the planet and life styles in cooperation with living beings, although this reading is a view from microbes, or can find a close relationship with the human body. This work is interesting to argue about the life of human beings that are competitive with all the implications that this entails in lifestyles and psychosomatic. Lynn Margulis y Dorion Sagan, Microcosmos. Cuatro mil millones de años de evolución desde nuestros ancestros microbianos, Barcelona, matatemas 39, Libros para pensar la ciencia, 317 pp.

an emotion that has become dominant in the body, via the combination of what the subject represent and feel. The mechanisms of construction are not different to their historical condition and their representations of a culture that places mobility and a way to work through a set of symbols, especially language. Thus, the realization of these aspects is a psychosomatic condition is subject to interpretation if we follow the path of an expression or suppression of emotion and feelings, which is personal history. People dump them somewhere, or more specifically in an organ that may be weaker or stronger on the inside of the body. It is possible to locate it and start working with it to extent that we can make it harmonize and not dominate the other organs, it means that other emotions are subject to an organ, such as slang, is a bully, and it is because it occurs in culture, nothing is free, is not random, is a complex process in the eyes of objectivism. Is it possible to achieve change? I think so, just have to do body work to change the memory it was built, that is, they may think it is feasible to transform the body from the recognition process development in a culture, and if we do this reading from the five elements of the logic of the Taoist, we can approach to this psychosomatic process and times as many deaths has caused unhappiness in the lives of Mexicans.

An epistemology of psychosomatic containing this reading of the body is in a position to find new interpretations of a phenomenon whose origin has been sought in the brain, emotions, and its effect on the endocrine, immune, that is, endorphins are responsible, requiring neuropeptides and neurotransmitters in the brain cells to create a system of stability so as not to depress the immune system. There is another reading of the emotions and feelings as autonomous entities that go directly to depressed immune system and lowering the production of lymphocytes or T cells. This reading is still fragmentary, the body is divided for better understand it, but the solution is made possible remedial. We can list the psychosomatic other readings that are in a different logic and approach to what we intend to propose.

Given the high levels of stress in industrial societies, the Germans and the Japanese have tried to find solutions to the organic type alterations whose origin is in the emotions. The proposal is to restore the homeostatic cycle; Hirai did an experiment with Buddhist monks and psychotic to try zazen meditation as a way to achieve the homeostatic cycle to be restored. Found that alpha waves appeared more rapidly in people who do meditation, and those newly incorporated this habit began to calm down when the waves appeared, which showed that therapy zazen meditation is a possibility to improve the quality of life and cope with stress. This leads to an attitude adjustment to make the body to produce some answers to a society that is becoming more competitive and destructive of human spaces and individuals.

The search for alternatives is ongoing; some authors have explored the energy fields to solve the problem of the psychosomatic. Drouot maintains that it is necessary to go to the search for new readings on healing; the therapies of the future, as he calls them, should cover the spiritual space[162]. But today the religions are challenged; it can be a utopia, a body that wanted to be perfect[163]. We then speak of spiritual development where the balance becomes a principle of inner peace.

Of course this is a more advanced proposal that includes deep work with the memory of the body to transform the culture established by and for the space of the family. A cross body by establishing a language symbols to express emotions can be the bearer of what happens in the process of international cooperation of the organs, but relations have been established are part of the need for security to the instituted in the bond of emotional and organic. An individual cannot be explained without these joints of the inner life of constant struggle with the outside or vice versa. This space, as a concept, is the microcosm of which we speak and is far from the fragmented vision. So what can be psychosomatic can

---

[162] Patrick, Drouot, *Sanación espiritual e inmortalidad. Las vías terapéuticas para el futuro.* Barcelona, Luciérnaga, 1994, p. 132.

[163] While society is struggling to institutionalize religious processes with a discipline that did not culminate in new ways of living life, control and discipline that culminated in new ways of living life, control and discipline appeared as instances when faced with the search for human perfection "to the chase: the body is the origin of inertia and sloth so: ignore it, then, and go about saving the soul, sex is dangerous: therefore propose celibacy as the ideal corrupt wealth: we must despise them, the power lends itself to great abuse: it is therefore better not to use it. In one word, the world is tempting: you have to run away from it. Raimon Panikkar, *El mundanal silencio*, Martínez Roca, 1999, p. 125.

be instituted in the body from the exaltation of an emotion that is grown in everyday life and manages to change the course of social security in mind, to become a conventional symptom or innovative in terms of pain or near expression refers. This means that the symptoms become more real, in terms of symptoms and signs, which means an illness or a condition that develops its ritual, its association, its significance, but it is not feasible to provide a timely remedy, it derived from the failures of medical or psychological therapies, because in both cases occur a conditionality or extent of symptoms. The patient's fear is an element that is allied with feelings or emotions that dominate both internally and externally, anywhere in the body. If you can change the biochemical or physiological processes with an emotion, it means that the road can be reversed, this is, an emotion can be diluted or lost dominance in the representation of the body. The road may be to achieve harmony in the internal process through meditation, breathing, and energetic balance among other possibilities and benefits for the body[164].

---

[164] Regarding the work of meditation, Khalsa says:
• Meditation creates a unique hypometabolic state in which metabolism is in a state deeper than sleep. During sleep, oxygen consumption fell by 8%, but lower during meditation between 10 and 20%.
• Meditation is the only activity that reduces blood lactate, a marker of stress and anxiety.
• The soothing Hormones, melatonin and serotonin are increased by meditation, while the stress hormone, cortisol, is reduced.
• Meditation has a profound effect on the three key indicators of aging: hearing, blood pressure and vision of nearby objects.
• People who meditate have a time consuming 80% less heart disease and 50% less cancer than those who do not meditate.
• Meditators secrete more hormones associated with youth, the DEA, than non-meditators. Men of forty-five who meditate have a measuring of 23% DEA more than those who do not meditate, and in women the difference is 47%. This helps reduce stress, increase memory, and preserve sexual function and weight control.
• 75% of those with insomnia get to sleep if they practice meditation.
•34% of people suffering chronic pain can be reduced significantly to start practicing meditation. Khalsa. D. Singh y Cameron Statu, *La meditación como medicina*, Barcelona, Diagonal, 2001, pp. 30-31.

To unravel the creation of diverse spaces inside the body is necessary to approach the experience to feel the body, or at least observe the process of possible diseases and energy imbalances that are focused on a symptom and are now a reality in the body, and must be interpreted with a new relationship of organs, emotions, culture, family, among other processes, to give a different meaning to the term *psychosomatic* and work with the body that gives opportunity to create other solutions.

9

The installation of a disease in the body:
some notes.

I

A condition is a complex body language, one answer could have various interpretations, too, is the ability to decipher the processes that have been installed on the individual. If we only consider the condition as a sign that the system is unbalanced in terms homeostatic, we are prevented from thinking that it is actually a more complex expression that has to do with social conditioning of individuality. A reading like this would be unilateral, and it would find what you want to find, with a definite pattern of symptoms and would end up the individual expression. In other words, this means that the patient continues with symptoms and even achieved the symptoms of his illness, grow, stagnate or become increasingly complex, settling in the body with an uncontrolled expansion, this means, the internal construction mechanisms are overwhelmed by an emotion, or suffering combined with the symptoms. The complexity of feeling on the subject crosses his personal story of suffering and ways to channel conditions, this means, which organ was installed by his pain, and if so was a temporary circumstance or a process of self-destruction. The implications of both bear fruit is about the ways the body deteriorates as the days pass. With regard to pain, increase it or make it more complex is a function of personal processeswhat do human relationships mean or even what does the subject want to do with his disease.

What results from this complex relationship is the individual process and the scope the subject must do in his personal construction: the complexity is substantial part of how a process is installed and starts walking, in the body, and has different implications, his expressions as they relate to the individual's culture, or his representations and meanings, family history and other social and individual processes.

The search for a solution to the suffering has become a demand for new ideas and proposals, epistemological, and above all, a conception of the body in excess of the established boundaries in society also requires a strategy that includes the concept of service to the body. It is a problem of epistemological, philosophical and even ethical, on what to do with the body and its implications on health and ways of living and dying. The depth of discussion restates the system of life of a body in society today. The excuse, are the conditions that may be of historical and generational type in a family, and that is not located in the enduring status of contemporary societies, the idea that things should be forever and the bodies want to be transformed into the proof. This ambition is carried to the search for enduring the eternal, with the understanding that human beings can be a certain way in the consumer societies and highly competitive, where they can find an inheritance of genetic and social. It is known that culture plays an important role in the process of building a human body: the ways to make life come together in the production of forms of life, and that is the starting point for the representation of an individual and institute what should be done with the descendants. The human being is capable of being built, is made by others, and in that sense we cannot say self-contained. Humans have a construction process, which involved not only the State, but also society in its inter-relationships and family. The particularity of the social subject, the individual states to choose: in some cases, compliance is established while in others cases he decides what to do with his body. This process can be traced in the family relationships, because through them are the ways an individual establishes his individuality; there are instituted tastes, flavors, gestures, postures, ways of eating, sleeping, and love among others.

An individual is constructed by a process that includes not only the homeostatic cycle; but also other elements such as the historical-social process, the family and individual level; this last on the list has other process involving the establishment, reorganization and internal network transformation of organs. The cooperation process to feed or get the nutrients from food is combined with their distribution to different organs and shed debris, but the chain does not end there, the network of organ function is articulated with the emotional processes of individual. An emotion can produce changes within the body and represent in forms of symptoms or conditions. This construction body-emotion has a horizontally reciprocal form, this means, the emotion can come first and then expressed in the body, or conversely, the organ and its expression can result in an emotional state. This internal process is not autonomous, is articulated with the forms and styles of life of an individual, so not everyone has the same form of relating their emotions with the organs; this means that the domain of an organ or an emotion is not predetermined. The process of building a condition help you understand what I am pointing. When a person gets angry can have in his body memory mechanisms to deal with anger: he is indifferent, gets diarrhea, painful bones, gives headache, it raises blood pressure, gets stomach cramps or the colon gets inflamated. The relationship can be direct or stored, which implies conditionality: anger equal to X demonstration. That response will be presented with their variants without any problems, but when the person has other implications in his body - for example if he is prone to a heart attack-, the consequences can be severe or death. An anger can modify the functioning of the pancreas can block the production of insulin, resulting in being diabetic, otherwise it can cause hypertension or future embolism or thrombosis, depending on levels of triglycerides and cholesterol. Emotions can affect the body, its relationship with the network operation may alter or prevent a person establishes his emotional state at will; when there is an internal growth ways of being and feeling into crisis with the changes and accommodations; hence the conditions are the expression of the internal changes in the cooperative network. The emotion and dimensions have profound implications regarding the ways that are installed in the body. The complexity of emotion and its path cannot be explained by the homeostatic cycle, or with the expression of the somatic or psychosomatic, because there is not a classification, but the realization of a process that goes forward and transforming from the time of installed in the body. Hence the complexity of a diagnosis that seeks to establish the limits of a condition. No need to walk this path to understand the process of a psychosomatic illness.

What we want to open is the boundary of a body process that does not end, but it can set the balance or be destroyed. The strategy is determined based on how we understand the dimension of the body in different logical explanation; the dimension of modernity and its implications in the body can not get away from education systems, information and power in close relationship with the religious development with cultures very important factor because it has become a mechanism that can establish a balance within the body. So it is not feasible a subject easily break with the moral that he regulates emotional processes and his principle of kill or love the other. In this logic, these aspects have an important role for a body to be moved inside. Religious repression produces certain diseases for which there is no need to talk now, but the delay in the villages is not related to the culture of the modern, if not to religion or ethics and morality of a society. Many diseases are linked to this process, but not all due to this religious mechanism, there is also a possible detailed mechanism, surely individuals who have built a pathology have been fond of id and that means that you can not detach easily, this relationship with the disease is a sure sign that establishing a way of living with the disease: it is normal and necessary for existence. Such an attitude not only exploits the family inheritance or socialized with other diseases, also leads to a destruction of the body that can be slow and deliberate so as not die violently, but does not occur in a linear fashion.

II

There is a boundary between the symbolic and the significance of what is meant by being human. A person is said can not go beyond what is established; establish patterns of representation can make the body does not move, establishing the form and the conditionality of its operation and no way to change the feelings, tastes or a lifestyle. This is when individuals come into conflict and contradiction in the process of how they feel or represent. The body rebels, opposed to the process of compliance and institution of life is based on oppression of freedom of being a subject can not be restricted, it is inherent in its natural condition, this means, a human being can respond to the circumstances and well beyond what has been said that life should be identifying the behavior of a highly competitive society. By contrast, some will be at a point such that only play the instituted. Humans may remain in that states without question, claiming that they are as represented but those symptoms or conditions are able to establish the boundaries of the possible with their body to survive in the concrete jungle.

The key is to discover how a subject establishes the mechanisms of disease in his body, how they arise cooperation networks between emotions and the organs and their expression of life, how to go beyond "I'm sick of something" and leave complaining. A disease is not single causal or multicausal, but a more complex process; is something that transcends what one thinks of it. There is a representation that involves a task with humans and sets the parameters of what is meant by symptom, so we must contemplate what to see in this search for psychosomatic illnesses and how they are established in the body, from where to make a reading that breaks with this representation leads to ways of feeling and building the disease or condition in ordinary life. The disease can also be seen as a necessary crisis experienced by the body, and even for some represents the opportunity to realize where they are failing, while for others it is part of a growing body according to where you live. The truth is that the crisis of a body is necessarily to see oneself from another logic.

III

It is comforting to think that life exists on this planet and we are one of the areas where there is even extending, life is only a possibility it exists or not, and it exudes a wide variety of choices to feel, to make the bodies and feelings have an independent life of existence; is something like an occupant of the body structure like the beings inhabiting a snail. Life and emotions are a challenge to learn, to feel and see what others have in their person. Maybe we should make life does not breed anything in the body, just do what you do and build feelings with no other hope to make a living with symbols. The symbols are set in the body and are growing in significance, branch out and make the outside world becomes an underworld in the body. A body that starts to live will not let anything dominates it, it is hungry and has needs that are leaving some gaps in the interior and this leads to the neatness of a feeling that does not contradicts in the flesh. Flesh that wants to be waiting for the natural world and emotion that matures faster than the body produce the difference and the body split in two: one conceptual and one flesh. The dispute of a space for both is a great dilemma that humans face with their crisis that are nothing more than the instant of an organ that is to accommodate.

Emotions are an entity to be taking over the system if there is no mediation. When a child is born, the breathing rate is not related to the principle of doing or getting something in the human relationship through the body. However, there are children still unborn and born who are expressed with a management of the body. The relationship built is in function of parental age and the process of stimulation that have been given; this relationship is established in their body before birth and see how the body will cease to be autonomous, only some functions can be set and then can be changed. This leaves a big gate to the exercise of free will, and this may be the room for a change, or, conversely, for the body to be ruled by emotions and lose its identity.

It is worth reflecting on the role of the brain and its relation to the organs. In recent times, have expressed various diseases of the brain that has a close relationship with the problem of the organs and emotions. Brain function is to apply nutrients to the organs, and it establishes the principle that one day the relationship to deteriorate and thus change organ function in the presence of an emotion inhibits the production of a protein. This way of relating leads to the cooperative network can be interrupted by the principles of competition or the domain of emotion in the body.

The deterioration of the brain is a constant in our time. Marilyn Ferguson's studies only show concern for a body that took over the body. Brain changes became an important source of research over the past 20 years, and so far the results are specified in opiate to relieve pain and new statistics for public policy. The brain can be read from different angles, but here it is dealt as a body that articulates with others from two perspectives: the nutrients and emotions. The conflict arises when an emotion is stored in the brain and competes with the emotions of other organs, resulting in the constant struggle within the body. We analyze the process of relationship of emotions and ways of building on the individual to determine how an emotion has damaged the body or brain. And this analysis should begin with a reading that includes the diversity of processes of construction of the society in which you live. That is, in a culture of resignation, competition, cooperation, anarchy, probably the effects on the body will be different with respect to brain function. The process of building internal relationships arises in the moment of conception of a being. The link with the mother and her culture articulates the structures of the different systems by establishing networks of cooperation necessary for life, but life for humans is much wider. What binds the fabric is not blood; it is not the interaction of an organ to another, but the dominance of the emotions in the body and the separation of emotion in relation to the organs. Autonomy can have emotions is the starting point for interpreting the reality of the body and its circumstances. The brain takes over the feelings and the physiological responses to or from their own representations, the sphincter control is a clear example of this, but the body is not only the brain and is not dissociated from it.

We have been taught to think with the brain, denying any possibility to feel with the body as if either were not articulated.

Becomes relevant where the process of appropriation of the body. We must learn to think also with it, that thought does not have absolute control. The reason, say the Buddhists it is a brain disease.

It is for this hegemony of reason that man of our time is divided into two bodies that inhabit the same space: the brain on one side and the other bodies struggling to keep their autonomy in relation to the life project. Life is not free of contradictions that foster new partnerships and mechanisms. The weight of years of building cooperative faces a new relationship within the body and from this perspective the sufferings of the past years, it may be the expression of a metabolism that has undergone constant changes to policies and ideologies that lead it to establish foster new networks and new posts within the body. All these changes also transform our ideas about life directing them to attitudes of fight, attack, defense, which only create fear and prevent us from thinking about other ways to channel our emotions feelings with all that implies in the way we relate to others. Moving a structure to be built in the body is not within days or months. Each generation inherits the following their ethical and moral codes determining a certain structure, society today also has instituted mechanisms but go beyond historical or sociological explanation.

A body immersed in a culture is born and this already implies a load inside, the process that will build its way of being and the articulation of its metabolism will be determined by how emotions are worked in the family space. The body not only expresses a dominance relationship between emotion and the body also includes the spiritual development of individuals is the most important in relation to the proposal of love by a space or for the planet, life is understood as a slogan to stand for inner peace.

This becomes more complex reading of the body, because although we assume the condition reading reflects not yet accept the fact that an emotion that inhabits the brain can control the body, or, conversely, that an emotion-hosted an organ can affect the brain to the point of ending it. However, some diseases such as Parkinson's disease, multiple sclerosis or chronic depression account for this possibility.

The body also has its wars: to the obstruction or denial of a particular protein, the organs are fighting nutrients and this contest can encourage disease gestation. This is not surprising since it is a reflection of the circumstances of our society.

A body not only obeys the principle of reproduction, but to build more complex processes is face by reason of internal decay and social view, that the outside is always an expression of the inner. Rationalism presents a body decomposes before the imposition of a single way to feel interior and exterior, and this rule of reason, already has outstanding accounts with the loss of harmony in nature and internal processes of the body.

The struggle that occurs in young bodies is related precisely with the construction of a response from a subject that has no roots in the dominant ideology, which means that organic processes are entering the process of adjusting a routine that escapes from the demands of a stressful and competitive society.

IV

When the body has no locks, an emotion travels freely within it and the risk of an abscess is unlikely. The human being is not only what he thinks, but also, what you feel and what you do with your emotions, since one hand is the thought and the other the way energy moves through emotion.

An emotion can change the course of the body and that is why we cannot speak of a separation between physiological processes and emotional.

Even the ways of processing the idea of doing with the body and its effects can be an important indicator of what makes an emotion installed in the body. The culture of objectivity also imposes certain emotions and beliefs to people, determine a certain style of feeling that has consequences.

In the process of building the body cannot be ignored culture as a factor in determining which of the human condition is what builds the emotions depositing them in a lifestyle. Understanding this part makes us realize that in a culture in which rationalism prevails, what exists in the background of an emotion is the idea of reason: this means the reason that we feel sets the style even when emotions are something that can not grasp the objectivism.

So understood, emotions leave us two ways: its social and cultural history, and the process of construction the subject is taking in its time and space. So we find people who do not have a defined relationship with his emotions, whose lifestyle can be very basic and even not in condition to compete or want what is within the reach of his hand. A human being, who is built or is understood in his style of living, is also predictable in his way of dying.

This prediction is linked to the ways in which he moves his inner world, the relevance he has, and how complex it is to decipher for the same subject. He can be scare to explore new emotions; can wake the fear of the unknown. Security gives him peace, but this lifestyle only makes the possibility of a mental block. It is built as a body that may not want or feel something that is out of the establishment. The credibility of what he feels and looks can be correlated with his body, but we must give a reading that permits to explore what can be felt.

An emotion can be many different things, but its effects on the body, whether it is endowed than one meaning or another, are similar. A reading may correspond to the language and lifestyle that the individual is immersed, which cohere and enable a culture or a State policy. This can give meaning to certain ways of defending the status quo or to interact with the family space. This seems very close credibility regarding what can be done with the life of his own and others', but that feeling that flows can lead to nationalism, love of family or couple, emotions all building and providing security and inner peace.

Humans are the only ones able to make sense of things, generating feelings of ownership. We transform love or any other feeling in something with a ritual, always based on our relationship with others. These feelings are installed in the body from the culture. There is another reason to make sense of an act as sexual, for example, surrounded by rituals and cultural performances.

Moreover emotion can also be part of heredity transmitted by parents, which affects the construction of a body and its relationship with the organs. Emotions are not subject to an election, we all have them, develop them and grow as part of a personal process that allows us to make sense of our relationships with others and with ourselves. Thus, emotions can be experienced only individually, since no body is equal to another, much less equal are relations built inside every body.

Thus, we have that emotions are integral to the subjects and, depending on time and cultural space in which we live, we will give priority to one over the other and we will build them in relation to others.

We can say that emotion is a source of disruption and changes in the body, to the extent that makes the body move in certain ways even when the subject is not even aware of it as only reacting to the circumstances of his time.

Emotion, then, is part of a cellular construction system which provided of new possibilities to the organisms as not only help determine the physiology, but also constitute a process itself that gives life and preserves it.

Just as an emotion can ill the body, other can save it. Everything depends on the meaning to be given to the problem of subject and his environment. It is not an object of this paper describe the emotions. We believe that all are valuable as long as they are directed to the construction of the body and maintain the harmony of the body.

We emphasize, however, the importance of deepening social and personal history of a body to look after this side of existence. Given that the body is composed of emotions, a human being which emotions are fractured does not have a long life or dignity, and this knowledge should be incorporated into any speech that truly pursue the preservation of life. Going even further, following this logic should raise to the level of indestructible value the care of certain natural emotions in children who later allow them to develop their personality. The call is necessary because we seem to have lost the course of humanization towards the technique, much progress can never replace the emotional side of life.

This very complex issue forces us, first, to rescue the history of emotions as a category that makes it possible to articulate the corporal process and social integration of people and on the other hand, develop an explanation that, while not absolute or unilateral, help us understand the process of building a society where emotion dominates. Thus relate this knowledge with mortality rates in the sufferings give us an indication of how emotions evolve in a given society.

These elements allow us to establish correlations between the pathologies of organs and diagnostic methods to detect those who have stopped working, without neglecting other readings that complement the knowledge of the body and its organs.

For all this we hold that an emotion must be built with a social purpose, accepting that lives in the body and is part of a cooperative process that allows the body's sense of belonging.

The emotions built into the body give us the pattern to give a deeper meaning to the exploration of new emotions. Up here, we've been following the less fortunate emotions such as - fear, sadness, anger, anxiety, to name a few - and results can be seen in the history of cultures.

Hence the importance to track the route of mexican society's history, in order to give sense to emotions that are destroying the Mexicans body.

V

The feelings, which as noted may result from a period and are combined with the complex demands of a way to exist. Also, humans can give way to transforming emotions. So we have desire that can become a passion, the passion becomes possession and possession becomes suffering, establishing parameters that prevent personal growth. A human being has the capacity to see from a distance, but what exists inside his head prevents him from doing so.

Emotions can even become protagonists in an action to halt the body as a resource to mobilize other emotions in other subjects. This means that they have the ability to produce movements, which in turn can generate movements in other bodies in a dynamic of subjection not so easily unlocked.

An emotion that paralyzes an entire family for example, part of a process in which he sets the tone of paralysis known to explore the emotions of others and force them to submit to his authority. The problem is that in this process emotions intersect forming an amalgam, which nullifies the intuition and prevents from seeing what is happening, allowing the emotions to win the battle.

A subject that has paralyzed any organ of his body and goes beyond the limits of his own control. Cannot control his body because it is conditioned to respond and work to the demands of others who can start from a need or a mere whim. Hence we consider that a body involved in emotions cannot be so free, because this process leads only to create new mechanisms of control over others.

The reason does not allow for an epistemology of emotions as described herein and that is why in many cases the diagnoses do not correspond to the reality experienced by the subjects.

However, this process is so complex that today will be much easier to assimilate into the future, because the body can not hide the effects of an emotion that is installed and demand an answer to a subject full of ideas about what should not do with his life.

Installation of emotions in the body, thus understood, is not the problem because it is a natural and social process. The real problem is how to develop this process and the social implications that this entails.

An emotion that spasms the gut or paralyzes the kidney is not an event free of cultural processes and in turn affects the social life.

This circular process, the way the body establishes a relationship between an emotion and a particular organ, sets the tone to make sense of the subject's existence, but to reach that point requires first identify what are the emotions that dominate certain organs determining the styles of living and dying today.

We then face a growing phenomenon that has only two possibilities of materialize, paralyze the body or make room for an internal struggle that will bring wear and suffering. The bodies of our time are a combination of both, in which emotions make suffer the body with opportunistic diseases.

This reality forces us to confront health systems, and of course the epistemology of the body. Emotions are now part of a market and a way of being responsive to social demands.

In a family, for example, or at a social gathering, all detect who is grumpy or cheerful, but not delve into the processes that led them to be. Question this way of life and history of a culture can shed light on this as it allows us to understand how the subject has appropriate ways of being of your family or social group, but only from work performed by the subject and his body that we can discover how it was that those emotions came to settle in it.

A reading of this kind makes possible to see other angles of the emotional component of a society and this allows us to think of new strategy for interpreting the lives of people and especially their emotional education.

Paradoxically, the analysis of this emotional education at the individual level necessarily leads to a social analysis that accompanies the course of the humanization of man, because a culture also receives influence from others and this influence brings emotions to be installed in that culture in the form of rituals and customs to intervene in the individual process.

We thus find that emotions are not as pure as one might assume they go through many processes, so it is not a unilateral look how we can address them. Only pride holds that premise, and pride in turn is nothing that the defense of an emotion and a sense of identity and attachment.

Learning from emotions should not invite fragmentation. By contrast, our proposal is to address them as part of the processes of a body facing a complexity that surpasses it.

We also propose from inside the body as a point of interpretation, we consider that is there and not outside, where are the answers.

VI

The logic of causality is, in itself, a way of living the body. It is about looking the immediate, what you have in front, but this way of explaining things did not come alone, but responds to a demand for the company to establish criteria of truth from what they see sense.

A human being who does not know the apprehension and despair can hardly meet the demands of a conventionalist society that rejects other ways of feeling than those instituted.

Emotions play then, as already stated, a social function to the extent that determine an individual's personality, but personality is something that can not be constructed or analyzed based on a single idea of the human.

Hence the importance of working with the body, which has been interpreted from all angles and which has made the most diverse theories that since its proposal fragment, just come to this spectrum that not only breathes and moves.

All these readings are animists, subjective, systematic, and moral or immediatists are the representation of a way of life the body that in turn corresponds to a particular time. The list is endless; from the theory attributed to specific characteristics you are looking for their DNA to unravel the truth.

In psychology, it is the discipline that interests us here, have made strenuous efforts to reach an accommodation that explains what happens inside the body symmetrical setting correlations only resolve the apparent level.

Psychologies constructed in this objectivist logic only give continuity to these other readings that leave no room for new interpretations to understand the complex processes that arise in humans.

We speak of a complex process that necessarily requires greater openness and at the same time, a certain scientific rigor. Psychoanalysis, for example, approached this search about what happens when an emotion, how expression of hysteria seizes the body, but the answer cannot remain only in the verbalization, the catharsis or the development of a metalanguage. The analysis of the unconscious through discourse is certainly a discovery that we cannot neglect, but can still be given a deeper meaning.

Other theories sought in the stimuli, but soon discovered its own limitations: the absence of stimulation, the physiological response is exhausted. So how can constrain the behavior, it is equally feasible to generate a new response from a different stimulus, establishing an autonomous movements endless cycle of conditioned responses, and so on.

This poverty in theories led them to combine any technical application and merge with alternative medicines without respecting its epistemology of origin, which explains its failure.

The aim should be, in any case, these theories provide a scientific and ignore both invasive techniques such as self-help manuals in such demand that reached in the stores.

Anything goes when it comes to making sense of what happens inside the body. No single theory provides a satisfactory answer, but at least we have the notion that the body is in crisis and to get him out is necessary to return to an idea of the integral and harmonious.

VII

History and anthropology are two possible ways to unravel the process of building a body condition in the reading of cultural development that both disciplines are in a certain time and space.

This is a reading that, without resorting to biological explanations, absolutist understands the body as a body crossed by culture, which gives students the body new arguments to give a deeper meaning to the subjective and symbolic.

This cultural reading, confronted with the current biological knowledge, would lead to a new position: neither biological nor cultural control today, the emotions give new meaning to the body.

The body-emotion relationship is the mode that gives new meaning to physiological processes and allows us to enter and understand and intervene more efficiently in the pathologies of our time.

The importance of emotional factors in newly diagnosed should not be ignored, because emotions are a reflection of the new process of social epidemiologies.

The truth is that an emotion is not visible in X-ray plate or under a microscope, and will not appear in an ultrasound, so we need new eyes to discover them.

10

The human body, culture and health

Introduction.

The body can not be something else, is what you have and in time and history is read from different views, so does the soul, spirit, consciousness, thought and mind. Have had a relationship with the body, whether as producer, the space where they live or production of something that is embodied in flesh or body, the complexity of explanations are diverse and multiple paths are the implications with users, academics, the philosophers, the debate over history is very valuable for understanding the current status of this body that is sick, that builds new problems, who ventures to build new networks of international cooperation to preserve the principle of life, is built from the culture and becomes a way of living it, that is lived and experienced body, the body becomes an accomplice of a family-style buildings raised by internal murmurs fanciful and not only is able to produce filtered water and feces highly sophisticated, it also produces ideas about itself, about others and is able to build his alter ego, the mind consciousness, ways of composing a poem, to paralyze the body in parts or in one of his members including the pain of phantom limbs; witnessing a process that actually is effective for thousands of years and comes with a whiff of confusion, psychological and philosophical explorations that seems to say that the truth of the body is something that can not be known by reflection or self-knowledge.

Years of wrangling over whether the soul or spirit living in the body, are part of it, or just a way to live temporarily and continues its path in the world of heaven or religion. We are-today-with a body more complex and full of new expressions, now has many resources to be stimulated in their five senses and make you develop only one or two and that makes a body type, ways to build it by humans, not nature's own biological barrier has been crossed and the historical-social have made a new face on the body, it starts from the dualists, represented by Platon, with the idea of body and soul.

The psychic body of Aristoteles[165] or Cartesian rationalism taking away the subjective and transforms it in analogy with the machine[166]. There followed philosophers like Locke[167], Heidegger, Schopenhauer, Nietzsche[168],,Husserl, Merleau-Ponty[169] and others, to show that philosophy

---

[165] "Consequently, for Platón the soul has an extra source and supra-body. Their union with the body is accidental punitive genuinely no intrinsic link. Be described as immortal. By contrast Aristotle's soul originates in the generation of man, the human compound of a compound "body", in which soul-form and matter are the essential elements. His relationship with the body is so intrinsic that the soul is "something" of the body because there is no human body without soul."

[166] On this matter there are plenty of texts that refer to this aspect of man-machine, I think it worth going to their meditations. Rábade Romeo says "Descartes, did not understand how hard that is to say, nothing of the human body. He did not understand more than the body as an extension. The human body as something that simply we do not have, but what we are, what we exist (pardon of the grammar) what we experience is totally absent from his consideration. "Op. Cit. p. 494.

[167] Locke, John, Compendio del ensayo sobre el entendimiento humano. Barcelona. Tecnos, Clásicos del pensamiento, 138. 1999,61 p. y Romanell Patrick, "Locke, un médico ecléctico". En humanitas Universidad Autónoma de Nuevo León, No. 10. 1969 anuario. 259-273.

[168] This text may well be an approach to the philosopher. Sagols, Lizbeth, "Nietzsche y el lenguaje del cuerpo" En Theoria. Revista del Colegio de Filosofía. No. 2 noviembre de 1995, UNAM.

[169] Battan, Ariela "La paradoja de la distinción. El problema del alma y cuerpo en Descartes y Merleau-Ponty". En: cuadernos de historia de la Filosofía.

working without the body has always lead to epistemological arguments that claim to be reality or a conceptual category and that can make a big difference between a concept and another on the body. But we can say that "man is his body" and from thence to understand the dimension that is said about the body[170].

The Judeo-Christian religion[171], among others, faced the problem by considering an obstacle the body to develop the soul or spirit, hence the need to punish it, hide it, submit it to devaluations in order to repress or deny the instinctive part of the body[172], says which is the animal part of our existence, so the body is related to sin, desires, drives and so it must be to prevent it consumes the soul. Saving the soul or spirit is the negation of the body[173]. In this sense, it is clear that any philosopher who forgets

---

Universidad Nacional Autónoma de México. Instituto de Investigaciones Filosóficas no. 3, otoño 2000. p. 13-15.

[170] These texts are a look on what is understood about the history of the body. Gilbert Paul. "Historia del cuerpo: expresión y libertad" En: Revista de Filosofía. UIA. Vol. 333, No. 97 Ene-Abr. 2000. An important aspect in the history of the body, the body of women. Should be understood that now we only speak of the body regardless of gender. But the author illustrates the theme "Gender relations are a key dimension to understand the female body as lived body. In analyzing the psychological levels and particularly philosophical, gender theory has indicated that women have their own way of living their body that is different from men and ignored by them." Godina, Herrera Celida, "La teoría de género en la perspectiva fenomenológica del cuerpo vivido" La lámpara de Diógenes Año, 2 Vol. II, Enero.-Junio 2001. Benemérita Universidad de Puebla, p.32. Certeau, Michell de. "Historia de los cuerpos" (Entrevista con Michell de Certeau) En: Historias y Grafía, Universidad Iberoamericana. No. 9, 1997, p. 11-17.

[171] A material that touches the subject of body and religion is: Durán, Norma. "La función del cuerpo en la constitución de la subjetividad cristiana". Historia y Grafía. Universidad Iberoamericana No. 9. 1997, p. 19-58.

[172] See Durán Amavizca Norma Delia. Cuerpo, Intuición y Razón. México, CEAPAC ediciones, 2004, 142 p. It is an excellent text that deals with critical insight and its confrontation with reason, even considering the use of the first to discover and develop the body. Open a door taboo for the rationalists.

[173] "As a liberation of the body is not going to be possible in life, the effort should be directed to make the body a submissive instrument of the soul is practicing the hegemonic character, by its divine nature, it is up [...] The

or denies the body away from the reality in which the body lives, there is the assertion Rábade Romeo.

> Our Western philosophy was born as (...) from a pretended reason, since its apparition, build a *universal rational* order. Obviously, such an order calling for a reason, somehow, had to be above the individuality[174].

This means that universal laws must be compatible with individual to the general explanation of the groups, therefore, the body is not the space to make such laws, so reason is the principle that unites, which can make sense a body that produces ideas and rationality; I think, therefore I am having success in that it becomes "a" truth "that no one can deny, we construct a knot that tightens as you ask questions, then the node becomes a vicious circle, only a detailed analysis at the source can give us an answer about those ups and downs of life that are built into the subterfuge of a philosophy that does not work with the body and only makes the body a hindrance, the weight of reason, will give a deep sense of search to the body, the analogies are part of a status that can not be built freely. Surely Descartes must be turning over in his grave when he learns that the reason is not only an entity that is the way, like a bird that migrates and binds the universe or God. It turns out that the reason and the body are constructed as entities of two levels; both emotional and physical. Talking to embody the reason or give a new face to the emotions can be thought of as an option but the problem is that the complex construction which makes the human being with his body requires us to go to a different logic to build the body and not the sense of social or cultural, no, I mean the conceptual building where the body can not be read with dualism or psychic life and even the purity of rationalism, you have to download or upload as applicable, the philosophers who consider the body as a philosophy and with it the possibility of what is

---

papers are clear: the body must "serve and be loved" and the soul "command and govern." Rábade Romeo, S. p. 467. Se puede consultar Pieper Josef. *El concepto de pecado*. Barcelona. Herder, 1986, 119 p.

[174] Rábade Romero, Sergio. Op. Cit. P 448.

lived and its implications for learning with and from the body, not only with reason, gaze, gestures, representations and meanings are part of this new view a body that is and becomes part of the world, is to turn the world than exists in any geographical area, new construction of any significance is what we can give to a body that can be released from the dualisms or the successes that the look of a reason is more than enough to understand that building now is not the same a thousand years ago, but it must be said, quietly strong and the body in its physiology has not changed[175], specialization of cells is the same for centuries, that means we can not make a body to stop producing insulin and nothing happens, it is not, meaning the body has been stimulated in their senses more than they should and that has created loss of inner harmony, it is fair to say that the construction of a body is not free of culture or food, as is the case of insulin or overweight, know that the body is in a new perspective on reading us to decipher the new codes are constructed and are not a problem of a philosophical, but they are, is the paradox, because each eye is a philosophical perspective, the attempt to decipher, understand the process of insulin by reason only gives us meaning that the machine is failing and you have to provide insulin and it seems the problem is resolved, but it is not so. And what I mean when I say that a body can not be the reason or thought equidistant from which can be explained with the purity of a thought that is not up to the circumstances of our times is, "the reason for the universal and necessary is a reason to be separated from the body and immunize it against the pitfalls of it[176]". No doubt that the body has developed his tricks, bait for empiricists, rationalists and confusion may be diverse and varied, one can say that potentiates the spirit, conscience, thought to the detriment of the body.

---

[175] José Enrique Campillo, A. *El mono obeso. La evolución humana y las enfermedades de la opulencia: diabetes, hipertensión, arteriosclerosis.* Barcelona, Drakontos, 2004, 235 p. This material is illustrating the physiological process of the body in relation to food habits and the effects of insulin and overweight.

[176] Rábade Romeo, S. Op. Cit. P. 448.

And chaos asunder of soul and body edge and be aggravated by the complete separation of two *complete substances*, of a fundamentally different nature: a *soul-thought-consciousness*, which is dynamic, active, pregnant by virtue of nativism, and against her inert *body-extension*. Its extreme substantial difference makes unviable causal intercommunication. As expected, knowing subjectivity leaves the side of thinking soul, to which attributes clear and distinct ideas, leaving the body relegated as mediator of obscure and confused ideas, and what is even worse, as an interfering element in the processes of knowledge, being necessary to guard against it[177].

This means that the body is not the resource to think or make sense to the actions of humanity, reason becomes the first domain of what is done on the planet. If we ask about the meaning of this reasoning and does not give space to the body in the art of thinking, we can consider that it is not up to what to do with the business of living; the body is the resource to explain man and his actions, said Maine de Biran, explain "its consciential dynamism." "The body does not need, the body hinders, and the body tries to dispense as much as possible[178]." The results are a cost that we are still paying in the field of health, high rates of psychological disorders and organic connection to the emotions, shows the mistake of denying the body despite the foolishness of some philosophers have gone so far as to "the idolatry of the body, idolatry that is gross when *body* becomes synonymous with matter in the most vagabond sense[179]." The implications involve a reading that does not see the experiential component, this means, and the lived body. Be in the culture, and even colloquial expression of feelings that are expressed in the face "I blushed, tense my gesture, frown[180] ..." it becomes dual use, which permeates the spaces of culture and the body becomes the space that is left of view and appeals to the idea of reason as the instrument to give a deeper meaning to existence, body aches are to rationalize or make them into something that is where

---

[177] Ibid., p. 452.
[178] Ibid., p. 455.
[179] Ibid., p. 457.
[180] Ibid., p. 459.

it should not be, therefore, punish it is a resource with double meaning; the denial of the body and what builds and produces. If we think that we inhabit is the result of the body, like, what we say, we understand that the dimensions of the discussion about whether or not certain the idea of a dualism that validates an action on the world and the body or make divisions that crave to study something that you should not but can be divided into any of the looks that say the philosophers, see what says Ortega and Gasset:

> The intrabody is colorless and has no well-defined shape, such as full bodied; it is not, in fact, a visual object. In contrast, consists of sensations of movement or tactile of the viscera and muscles, for the impression of the expansion and contraction of the vessels, the minute perceptions of the course of the blood in veins and arteries, by the sensations of pain and pleasure, and so on, etc. Our mental life and our outer world are both mounted on the internal image of our body that we carry with us always and is like the framework within which everything seems ... carrying each his intrabody with an enduring company, we do not usually stop to think about it. It is the invariable character involved in all the scenes of our life, and, therefore, does not bring attention[181].

This means that trying to show that there is no division between the outer and inner are one possibility in the body nearly to the idea: what exists outside is what exists inside on different dimensions, but are part of a reality which builds the body and the process of intention given on the inside in relation to a principle of harmony with the living space, but is trying to show that the divisions may be absurd while "is the invariable character involved in all the scenes of our lives "one wonders who can live without it.

It shows the other divisions that are used to give an order to that idea of what should be addressed to the understanding of the body:

---

[181] Ortega y Gasset, J. *Vitalidad, alma, espíritu.* Obras completas II, pp. 456-457. En Rábade Romeo. Op.cit.

First: by body we mean a physical or physiological organism with specific structures: skeleton, nervous system, arterial, venous, and so on. The body thus understood is a physical object of special features, like ultimately to those of other complex living beings. In such body can do scientific studies of various kinds, submit it to quantification, analyze it in different ways, in a word *objectify* from plural perspectives.

Second: by body we understand also that indefinable we feel, we experience, experience in almost total immediacy, something we can not *objectify*, because, to objectify it will no longer be the body, immediately experienced[182].

If the discussion of the body can polarize these two trends about what is meant by the body and brings a background in particular the idea of man, the chances are not separate due to high-risk, leading to approaches that cross with the soul, with reason, with the spirit and especially the ways of explain the how human beings live: it may be a disguised dualism of pure science or science that embraces the culture to understand so big event; the body and its expressions as complex and without giving an absolute idea of what it means new object of knowledge, including the body and the possibility of knowing the immediate world and its own world, that really feels like inside reality, experienced or intrabody.

While it is true that man is the body he has and there are two possible readings; what is objective and what is lived or felt. Then maybe it is time to seek other readings or other schemes to understand our reality, our problems, in our case: the psychosomatic ailments. Expressions and manifestations of the body have become the height to give a new meaning, so we had to see what philosophers say about the body and surprise found a rich discussion on the subject and there are arguments for giving and giving defend to a position or another and this diversity, however as rich as it could be it does not explain what we live in the reality of our time, how to explain and resolve the problem of a health condition in the colon, kidney or other organ. Someone can tell me that is not subject of philosophy or epistemological philosophy. I regret to say that it is because

---

[182] Rábade Romeo, S. Op. Cit. 460.

of their field or reading to understand the body, man; reality is constructed in any immediate geography, if the principle is explained the reality of man and if man is body without more, then we must take the point in its various expressions, borders and fragmentation are chimeras that can not withstand any analysis impose them in this new reality that brings us in the microcosm that is the human body, it is clear that we were not introduced as reality is constructed in that geographic reality of social space and the readiness with which the social and cultural progress as slow responses or other temporality in the body can respond or adapt to the times that are in demand socially and this point is perhaps not considered in this study the body as a product and producer of a time and place we must not see him as reconceptualize and mass accumulation and history that coalesce into a crusade for society and moral beliefs of sin, culture, every society has its eyes and his explanations of the relationship of emotions and organs that process is what interests us, how is structured to give us a new style in the body, according to the reading of Western philosophy that aims to give pure reason and buried the body as a burden that does not allow the purity of the soul or reason and that the body is just the space that allows this conjugation process, the reality is amorphous and produced by the body gets confused with reason or the other self expressions that consciousness becomes the common denominator of the body changes, but it can not make sense the increasingly complex expressions of a body that responds and is beyond the brain dialogue, you can not only make sense with the rationality that breaks it, it must go through the junction in the body itself, for that we need not only see what is done in the immediate past of any geography, no, I must be a new epistemological principle with the concept of the corporal and the body. We found that philosophers such as Aristoteles used the term of elements and their construction process in the body, mergers that occur externally and opportunities to understand that other process that arises within, but not elaborated, the logic of gnoseological type fails the form, so it can not go beyond the four elements such as fire, water, earth and air and does not articulate with the body in terms of the relationship with the body process and emotions that make sense in a body and a society, are: "The world, the human world-is what it is because my body-the human body has an specific team-organ receptors that necessarily screened stimuli and impressions I receive. It means that if you change our receiving equipment automatically changes our world[183]." But change does not only think that it changes with possibilities of evolution, we certainly need to discuss the construction process that makes the individual in a society in that it is a body and is a world that moves, has its own life, conditionalities are part of a process that encases or set it and it does not let you look like the interior, in a sense the use of recipients of the senses that are related to the organs are not a natural operation, specialization or the subtleties of their receptor functions are articulated with the inner world and its connection with outer space, we can not conceive of a foreign body in this outer space that wants to impose or condition feeling styles to digestion times, so recipients have to specialize with a correlation with the intentions of an individuality that can choose or do complex processes inside. The management of the information or the stimulation it receives the body is not up to a process of decoding or decryption of the new codes, now we can say it is very true for a body.

The combination of the social world and the senses makes a selection and a culture of education for all senses, this high specialization may vary with the corporal world of a subject, this means, there is the choice of a personal world that "we can intersubjective choice "to give the individual nuance" The central idea is that the body becomes a "here" all decided by the *presence* of a *contour* which lies in the *circumstantial* nature converged distribution of his many moments of presence. Go where you want, the body is still "here" that calls surrounding the appearance of all objects it perceives. And the circumstantial nature of that boundary, its distribution in peripheral areas of gradual withdrawal is established by this face scheme whose key is the coincidence of the *presence-attendance* of own body, *witnessed the presence* of things perceived. We would say that this coincidence is at its maximum density in the "here" of the organism itself and diluted in spread radical expansion, in the presence of the objects perceived[184].

---

[183] Rábade Romeo, S. Op. cit. 589.
[184] F. Montero Moliner, G. del Toro, Madrid, 1981, p. 145-145. Citado en Rábade Romeo, pp. 593-594.

The here and now makes sense in this statement, we can not go beyond the existence of a here as a body, that is the reality of a body that is inserted in a temporality, a countour that impose objects, which affects, that produces a look at the body itself, so here becomes the starting point for our case, with the understanding that it is the body that has a presence and becomes a space experienced here, become flesh that articulates with the organs and emotions in the space of the body which combines the processes of a here that links to a reality which is the realization of a story, so the body is part of this world and the world at its most thin and construction of new needs to give another meaning to here, this means we can not let the body is only the presence of an object made of objects, you can not talk as if it was a victim or the result of what is in its environment, the expression of sense and the connection with the operation of the organs only shows us that here becomes the expression of any physical reality of individuals, so we can better understand the process of a body that is done, it is built, it chooses, it transforms, which will rise to the sensations and decodes or returns to the others, the body is also the here and now of internal derangement of the organs, we can not conceive of this process the removal of a symptom which is expressed as the other reality of the here and now, that is not the reality that is constructed that allows concepts to make sense of suffering is the space of an inner process of the relationship of elements and organs that are built in a space that is defined as the body and its space-time with a culture with their emotions, their feelings and ways of expressing the joys and tears, it is not possible to speak of a body without the here and now of the body's processes, it is not the same one hour of pain than joy. "In a word, is because our bodies perceive experiential mode radically succession, timing and rhythm of time[185]." And that is present in the psychosomatic process of people, which means that time plays an important role in the individual to build his condition; either with a long or short time does not matter, the process occurs within the subject and will give a different look to his suffering and that's just the beginning of his feeling, is part of an inner reality that technology provides and the subject can be built with his rhythms, this implies that the way to live and time is not the same,

---

[185] Rábade Romeo, S. Op. Cit. P. 596.

so the experiential component into the body, provides a way to build here and establishing different movements to other bodies who are with another temporality and thus also the interior can understand the logic of what people do and what they make in an immediate representation of reality, that's the point; how psychosomatic can be explained from a reading of the phenomenal and can make sense of the gnoseological perspective of knowledge against a body spaces and makes movements to decipher or articulate this process with the health body is not far from the significance of what we understand about the health of a body that is not only physiology or symptoms that express a reading code that does not give to other reading-the psychosomatic- considered different, it is the domain space, with the legacy of Descartes-man-machine, of a reading of corporal machine type and the new corporal realities are not in that look[186]. This looming reality-is-in individuals can not be studied for their understanding and solution, from the concept of absolute space and time, one can consider the time and space of organs and the relationship established with the elements as wood, fire, earth, metal and water, it helps to understand the process of a new feeling in individuals, with the understanding that times sometimes do not tie in the individual, of course it is not unreasonable, recourse to the process of theory of the five Taoist

---

[186] The establishment of a logic that is held on the use of reason makes us see normal lifestyle, which does not allow other possibilities of construction of the subject. Whatever the lifestyle, it may be a manual or intellectual worker they do not have many options for living their body freely; the living process in the society of their time were emotional and physiological alterations that do not give an opportunity to be them. A man who works for compliance cannot be him, that means that psychosomatic responses are a shelter or a possible answer to survive, we can say that creativity cannot be sublimated. Diseases of the workers are not free, are the result of a workplace culture, geography, nutrition, etc., But the interesting thing for our time is the loss of a humanized notion that leaves no options to have a decent quality of life without certainly live well is to lose the identity of the expertise and relationship with a product, an object, it throws us individuals who do not find their true self. Who am I? It is a question that cannot be answered in that circumstance. An individual and a worker: they are two different concepts that we have to see with the idea of the general concepts of work to go into the background of the psychosomatic process in the labor sector.

elements of the proposal is not different to this world body here and now, which means we're not leaving without history or culture and make a new beginning of new meaning to give another twist to what is built with human beings in our time and do not worry about what to look for in the body, I think the spectrum of reading beyond the reading animists or those which say the reason is the joint space or denial body, we should use psychology in a process that sets boundaries on the brain or consciousness, even in the pursuit of neuronal connections in the process, the new reality of the body may exceed these explanations or alternatively can be enriched to give a different meaning in this complex process is taking place inside the body, breaking with the Cartesian principle is substantial and learn what the body does is not loose part of an isolated principle as a process that can not be viewed in relation to other processes, new physical reality is converted into a sea of demonstrations and fishermen are using troll with various baits to fish an idea or an outburst of the body to say this is what happens, but the process is more complex, perhaps the fishing nets are not enough, we must see it in the sea of his story or think of it as a principle that is not free at all, is part and is the world of the here and now.

It is inconceivable to talk of the body without subjectivity, without emotion, without the organs, except without the culture and nutrition, the reality of a moral and ethical crosses we can give an expression to the organs and sense of the bodies which are in a time regarded as the here and now the culmination of a continuing story, no purpose in the body, only expression of the past and the present in a combination of feelings and colors, tastes, smells and touches to give a new face to the long wait is it that the leaders have done their body and so is the body of the governed, social spaces are not free of conditionalities and less of the implications they have on the body those not yet born.

In short 1) the body can be considered as a possibility to change the ways of the pedagogy carries with him today. That is, compliance is a limiting factor to make sense of what I say, if we understand that education of the body is only the social reproduction there is no possibility to alter the principle of the body that is submitted and only makes an extension, which means it becomes nature. It looks natural all what happens to us, the lessons are just a way of seeing the world with a conviction that that's life, even so must die and so must suffer, humans become carriers of education of future generations. This pedagogy is not taught in the classroom alone, are the lifestyles, are the ways to feel or to die than become institutionalized and become second nature to all eyes. One hopes the teaching pedagogy or death of parents to say so will ours.

The issue is that there can be and this is where we enter into the background 2) the other view that we can show, tells us: what we have in the body can build a basic principle of election. The body is the space where we can build with the election, but that's not so simple. The processes that occur inside the body have been denied, it is predominantly a rationality that does not allow construction of new possibilities, dogma or denying what you feel is the cornerstone of a culture that does not allow the body to build with all we seem to come from outside and inside is set, the search is done on external sites and is not intended in the body that has its pedagogy, its construction process that is sometimes good to look at its collection process to build the body you want. Exercises to meet, make the individual can awake is a principle that is held in this logic of a body as space allows to make sense of what fits into your physiological and emotional process. The new pedagogy of the body is the expression of the technology, is ancient, is the same process that has existed for over 30,000 years, this means that the education of today silenced the voices of the body, from feeling or build new processes inside and outside, the body of today is not part of the new, is the denial of its interior, its space is filled with things that are useless in everyday life practices, the body is not what we have said, is more complex, is the space where you can build new networks of cooperation, where you can feel and express what you want, go beyond the finished expression of meaning, this means giving another sense to the process of conceiving the body as the place to reproduction.

Attended as an event that requires us to give another meaning to that interior that seems distant, as Descartes put the doubt, that Spinosa gave no other way but did not agreed, that the teachers of our days have a body and do not know how to educate it, also this is true, but it is also false because they say they have been educated according to the established canons, the truth is that with a "good and bad education equation" subjects become in something that is not expected of them; they are the antithesis, so one wonders where is the innovation process in the education of the body if not working with him, this leads to a more complex depth about what it takes to see the body or feel the body. Well we can go to the retreats of any practice and feel that something happened to us, but it only moved, it was not, illusion and deception makes us play on the body, the ego can make us believe we are good at what we do, the look does not go beyond a simple feeling that transcends the body of the establishment. The discussion is open, the rest time will tell.

III

The explanation and solution

11

How can travel an emotion in the body?

The human body can be overwhelmed by an emotion. Upon entering a phase of tension or stress may develop defense mechanisms to preserve the principle of life as result of cooperation inside and outside.

World cultures built their emotions in relation to their work processes, organization and the social division of it. The rituals, ethics and morals help to build ways of feeling and get excited. Emotions are not unique to capitalist societies or the undeveloped or even socialist, are part of the human condition, where there are individuals; emotions are the expression of its time and geography. And in the space inhabited by the emotions grow and can have the domain in the human body.

Emotions conceptualized as a unit that articulates with the body's organs are a complex set in its internal relations with other ogans and other emotions, in a cycle that has a basic principle: to keep alive the body using two resources: the emotions and functions of the organs.

It is difficult to understand this process if it is not an approximation to the emotions, but with the logic of the unit, it is conceivable a body without organs and emotionless as little without a history, a culture, a type of work, geography and a specified feeding. Fragmentation is a ghost that is dissipated in this nosology of emotions.

In the ways that different cultures organize the emotional life are human relationships, which in turn reflect the process of articulating the feelings and religion. This means that you cannot generalize an emotion in the world, except if the mechanisms are instituted to obey the wishes of those who want to dominate or subjugate the other. Well may we say that there is ethnology of emotions[187].

The journey of an emotion in the body is not straightforward, due to a process that becomes complicated according to the diversity of options that has built the subject in his personal history in his intersubjective relations and his relationship with the family. The construction process is not only part of a compliance or social determinism, but come into play a multiplicity of factors that need not be defined because it is constituted according to the historical status of the society in which we live. The process of creating the subject is what makes the difference, since there is no rule for the trip by an emotion; it has its share if it is made from the brain or if it occurs in a relationship with others. This means that the body will respond with their ancestral survival mechanism in a situation of danger, but now is not the same answer to climb a tree, running, jumping, etc. Release adrenaline and answers are also afraid to defend the body, life, involving the rational development of strategies to defeat an enemy that is hypothetically threatening the life of the subject. Witnessing a new process not only documented the life history interviews, but we also have to study and know what the route of the trip, according to the thrill in the body and sustaining us in the process of finding the relationship they are developing within the acupuncture meridians[188].

Follow your path puts us in an important point; we know there is an organization of the meridians and the course of emotion is occurred within the same network of meridians that allows us to give a different meaning to the ways in which travels an emotion, so we must take into account some circumstances in which it develops the event that triggers the emotion: when the event happens, the season to make a correlation with

---

[187] "In contemporary terminology, ethnography is the effort to develop rigorous explanations and scientifically cultural phenomena by comparing and contrasting many cultures. Instead, ethnology is the systematic description of a single contemporary culture often for fieldwork. Thomas Barfield, *Diccionario de antropología*, Buenos Aires, Siglo XXI, 2000, p. 210.

[188] Acupuncture is a treatment method by inserting needles into the body to balance energy meridians, which is the network where it circulates the body's vital energy, called *ki*.

the meridian that dominates. This does not mean that emotion is something that is encapsulated or has a presence among the body spaces.

An organ and emotion are an existing unit in the building process of human beings, the question is how it is establishing in the body, how it moves according to the processes of human relationships. That is, the meanings and representations in individuals play an important role to understand this we insure. We can see how a body facing different levels of tension and commitments moves this tension to the shoulders, the stomach or colon, and becomes a headache so forth.

The displacement of a contained emotion shall be in accordance with the levels of anxiety and the process will become more complex if it is added to family or social pressures. The body will have no space of tranquility and your dreams will be disturbed, organs, then, exhibit disturbances and imbalances, as the years pass, will become chronic and will move according to the work the individual will develop or not in his body[189].

In an emotion in these circumstances is not seen the beginning or end of the process that has happened, and ends up becoming chronic and degenerative in the subject; the representations and meanings of itself will be the person to neglect or the excess of care that will lead to solutions in any space which will provide the cure. The truth is that individuals make long journeys seeking a cure and make a lot of drugs to relieve their discomfort. Unfortunately for him, the journey of emotion can not be stopped with medication, rather it will leave sequelae in some organs (such as the painkillers) is not possible to track the journey of an emotion with x-rays or an ultrasound; this journey can continue in a silent and asymptomatic manner to the patient, relentless rationality of the person who wants to explain his condition indicates that he has nothing wrong.

The nosology explaining a disease characterized by symptoms impose on the patient so that he means what he says and what develops as the expert said. Thus it is feasible to build a patient symptomatology that does

---

[189] I refer in particular to the work release of energy in areas of the body caused by hatred, claims and grievances; this work will allow a good transit of emotions.

not correspond with the actual process that is living as it is much more complex, as the travel route of an emotion in the body.

Approach this idea of emotion can be complex in light of the current medical understanding. Those who do not want to see with these eyes have decided to seek a different route, it may be with an exogenous or endogenous. Knowing how to trigger an emotion can be useful to plot a route from the theory of five elements, allows, in turn, meet the internal networks of the meridians that have to give a course to these emotions in their process installation in the body; not a harmony that is installed, but a disharmony that will become a new organization structure in the body. From there we can reason an interpretation that accounts for the possible whys of symptoms that are not located in a particular organ but is the expression of the travel route, which is causing problems in different areas of the body. This is a process that merits different eyes and, of course, a different conception of body and emotion that goes beyond the concept of being solely the result of a chemical process[190], the logic must be different: a chemical effect, makes movement connection between the glands and causes the body to activate or retract. Its effects are mixed because there is no single route.

And diversity is a function of history and geography of the body and the family cultural heritage that gives the individual status and that lets him move his interior feature. The history of emotions in the family is related to a society that has been educated in one way or specific forms; implies that individuals may be part of a process that does not articulate with what others want; that is, to build ways of being and feeling, the meanings make sense in the body; such as rude is not widespread effects on people, but everyone knows what it means.

A human being only has the burden of emotions according to his organs, because the construction process that makes his body gives certain place to an organ, which can be dominant. Of course it is possible to see the logic of the relationship established with parents; they may represent a key organ in the family, usually the father is liver and the mother heart or kidney. Of conjugation or election made by the subject, of these two

---

[190] Antonio Damasio, among others authors, argues that the process of emotion is the result of a chemical product.

influences will depend his personal process, which will allow him to say that he is more a type of emotion and organ that another.

So the excitement should be explained in terms of family history and the different processes that the subject has lived, this is, what is sick throughout his life, it is clear that the set up and travel, the excitement can generate only shredding or discomfort, depending on how long the subject is living, that process, and, above all, what he does to work it. The process by which the subject builds its symptoms, is part of this course, so the treatments can not stop the journey of emotion, but less cure, however, distract cheat the system or even block it so he does not feel and meanwhile the excitement continues its route. A treatment may temporarily releave the discomfort of the subject, who will believe this is a cure, but this feeling can last a few months or just a few days. That's why we can not cure gastritis that is the result of an emotion, because what is generated has not been resolved, whether such colitis is seen as a process interwoven of emotions usually adjective as "nervous" (Irritable Bowel Syndrome) so that is sent to the box of "unsolvable".

A body living this process will send messages trying to be understood, but often the subject does not know or can not hear them, the drugs have anesthetized, so it has the culture of "no pain and no feeling." This seems to be a part that denies human beings from educated rationality.

This set of explanations are a conundrum for public health because disease rates do not decrease, even with advanced technology; surgery may be increased, but people report that they continue to feel bad, which means that maybe we should make some adjustments interpretations of the implications of the concept of disease from the emotions. We assume that a body is a microcosm where emotions are characteristic of an internal relation in the body, and accept that when it domains, it begins to wreak havoc in the organs taking a toll of dominance and autonomy, and destroying relationships harmony. Let us explain the idea: an organ is an emotion correspondence which in turn is related to another organ in three types of relationship: The creative cycle is one of generation, the opposite force is the control cycle and tonification cycle, where the Earth element is considered to be the fulcrum during periods of transition. They show all the possible ways to connect and especially highlights the complexity levels established in the body, which means that not only is a body because it corresponds to each organ parts of the body, this means, may be involved joints, eyes, nose, hair, tongue, etc., which at the time of analysis gives us a clearer idea about which body is the one affected by the route emotion.

However, we can add other items such as food, work and sexuality that lived in the excess (or absence) will have implications in certain organs. This shows a different process in the life of a subject because the mix of dominant emotions will not correspond to the affected organ, this means, exalts emotion and it can rule on the subject. This applies, for example, a thin person can be angry or passive kind and his actions may be changing, or is that governed by the emotions affect the body in question because it minimizes, that prevents the body from growing or become the repository of this emotion, of course, end up involving other organs.

The travel route of the emotion is not very complex because it is stereotyped and stigmatized becoming predictable to the subject. Perhaps with this type of case we will not have many problems, but in other cases, where the subject elaborates rationally, may represent a travel route that prevents to give a precise location, changes of place because the subject moves it; in which case the wear power is considerable because there is not stability in the process and excitement, which was dominant, happens to be a traveler who crosses paths with other organs, then we have combined symptoms: can present colitis, swollen feet, diabetes, temporary deafness, urinary tract infections, digestive disorders, etc., passing from one discomfort to another, giving the impression that you are always sick. It reads that the subject has an inadequate management of emotions.

The path of an emotion, assuming that is exogenous, start with the stomach and cross the meridian that is in progress at that time, will make its journey through the other channels or, failing that, will stay in a body and cause severe health damage, here are illustrative of the heart, liver, pancreas, because the effects are deadly or perhaps geste a chronic degenerative disease. Hence the importance of identifying the process to stop the travel route or change its course. When an emotion is endogenous can be related to stress or anxiety, revenge, hatred, spite, envy, jealousy, etc., Is a process that builds from inside subject, so you can see that the effects will be medium or long term; the path of the emotion is silent, and while is not alarm one day become crises. This emotion is installed in an organ or body part, even get them to stop producing certain protein or prevent the operation of any gland effects in some other area. Emotion is a profound expression of an attitude kept in the body and manifests in different ways. That is why the changes generated are considered part of an irreversible process, but can be worked with the travel route to stop or change course.

The complexity of this relationship emotional with endocrine system cannot be understood without reading the meridians; usually the subject keeps emotions of destruction and death, information that can go from one meridian to another until it is installed in place where it can destroy the body. The most illustrative case is deforming rheumatoid arthritis.

Build this nosology of emotion leads us to consider whether they have autonomy in the body or not, we believe that yes and no, because the construction process that some people have in the lifestyle and make their emotions be part of a playful state or violence. Build a lifestyle that involves only competition can minimize some emotions, however, are excited when an individual achieves a variety of states with his body; the ignorance of the consequences this has on the body may be the reason although we understand that this is not enough in this field of health because the body has become the space in which are deposited a number of actions that can trigger or enhance the emotions, whose implications are only at the end when the subject died of a heart attack or diabetes, etc. An emotion today is as important as an organ, reason why we should not make fragmentations and make the body become in the living space where the total emotions reside, and work with it to see it as a logic entire process, and health may be differentiated in its complexity with the emotions or the organ. An example of this is when you get an infection and it is attacked with an antibiotic and it gives way, otherwise it is when the suffering does not stop with medication, then it becomes necessary to think of another disease process, because it is no longer possible to work concepts of agent, host and environment.

So our reading approaches the process of construction of emotion, where it is necessary to understand it, since knowing the trip we can build dams, or give methods to unlock it and exit from it, because it is clear that the body can not be the depositary, for much of its integration cycle or control cycle. The implications of this work against the principle of cooperation in the body, so immunosuppression plays an important role in this process. Change the information that the individual constructs conscious or unconscious gives us clues to develop a strategy to make sense of why patients report feeling that they are moving in the stomach or other body parts. Cooperation is a concept that is best understood if we think that the origin of life is in his cell and in the memories and emotional; construction of this international cooperation networks are the basic principle that allows the ride of emotion. We have located these networks via the acupuncture meridians, the intersection points of these meridians inside the body are part of the key to understand that an emotion is not only a feeling or some secretion in the body, but it moves by it and has, of course, a particular way in which progress, stopping or moving forward according to what the subject does in social terms. A body is not as free as we sometimes believe as it is linked to emotions and physiological processes, with the status of his personal history; these are areas where our interest lies to learn how this pathway is linked into the microcosm and also with the macrocosm. We should add that today we could say it is a self-regulating mechanism in the body that can keep a certain balance, against adverse conditions.

II

The treatment of emotions generally subscribe to a logic of psychology that can work them from the removal, elimination, relaxation or verbalization, or from the exercise of muscular tension and relaxation, yet may meet again after a time, this means, it has a relapse, which indicates that the treatment worked only as a palliative. By not touching the body's emotional memory of the individual, most likely will have the same symptoms after having downloaded the tension that produced the emotion, thus creating a vicious circle. This idea of permanence of the emotion in the body is directed to the integration cycle and sometimes the control cycle excercised by emotions, either with different emotion or a particular organ. Nor are we saying that the emotion must be eliminated from the body that would be a mistake, but a balance must be kept between emotions. Its displacement in the body only generates a new process that can not be explained with the clinical eye of orthodox psychology with the understanding that starts from fragmentation of the physical and emotional, but in our case we have new questions about the point: how is that an emotion can destroy organ tissue and even cause death? We can accept that psychosomatic processes are aged in the human condition, but the rationality raises a single reading of the body does not accommodate this new understanding of emotions, arguing that they are not objective.

I do not know what the criteria is to support this idea, for objectivity, we know, is a construct that allows the eye condition; which is why it may be that objectivity is only one meaning of reality that is represented by an eye full of cultural loading, which means that man is the product of social construction of space and time. If we abide by this idea we can say that emotions are product of a social historical process, this means, that individuals have their share of choice, whether consciously or unconsciously, the predominance of emotions in a family can only be understood by the family history and the work space. Emotion is not reason for discussions to take the demonstration level, its existence is part of the human condition, even the arguments about which ones are prevalent in this part of the research today[191].

However, we try to go to the more immediate subject, the process of change in emotion, but sometimes there is no change after several years of therapy. That condition must tell us something will have been built shelters that are part of a process of complicity and simulations of what is human or indeed other options are in the process of therapy as long-lived? If so, where are the solutions and options to our time, and what we look at the new university environment that is giving options to new social and individual constructions of society today? Are certain constructions only justify the budgets of the institutions that have entrenched samples providing therapy and medication? It may be the boundary of a feasible explanation that only find a way back to the same thing, or is it takes to

---

[191] Among others may see: Harold Bloom, *La invención de lo humano*, Barcelona, Anagrama, 2004. Y Roger Bartra, *El duelo de los ángeles. Locura sublime, tedio y melancolía en el pensamiento moderno*, Valencia, Pre-textos, 2004.

break with the established to give a new reading and a new alternative to human beings who die with their emotions unleashed in his body, which it seems you can not see when they have to stop? These questions are for psychology and health workers, since statistics are increasing, especially mental disorders, suicide, chronic depression and sadness. Disorders that are silent because of social embarrassment makes you hide, you cannot see, but the statistics do not stop lying about this new reality of the inhabitants of large cities.

This new physical reality has shown us the limits of existing psychological models and, in particular, has led to the rejection of a rule and the dogma of the fragmentary that does not support a different perspective on the psychological phenomenon in its various demonstrations and constructions, but that inevitably culminates in premature death or a chronic degenerative disease. In this new reality you have chosen to call "the psychosomatic" in spite of the term existed since the early XIX century; well worth redefine it with another concept of the body and psychological; suggest that reading is part of a more complex process which includes the concept of microcosm of the human body.

A body that has faced the burden of history cannot escape free of charges to be instituted conditions in a culture; social memory, we can say, is incarnated for generations and is modeled bodies becoming natural and normal. The most illustrative cases are the marginalized communities of official history, leading a dynamic, a time rate and space in which it is hoped those new generations do as parents and grandparents did. This building process does not exclude the emotional processes and the prevalence of body changes, that is, they will die prematurely and have no choice in terms of generations, his emotions have to be a distinct feature, which means to articulate with the size of the organs, which is sometimes supplemented with a consciousness that can carry the body to the extreme. The construction process of the bodies may be stopped or taken to extremes, especially if we accept that many of them are not fed correctly and the organs are not developed properly; so emotions can be in this dimension, same thing happens to the soul and spirit in the body. The complexity of the election is full of fears and doubts, a complex process if we want to generalize. From this stems our claim that each body has its history and, of course, every social group, which means that emotions have to take a route different from others, because culture plays an important role in emotions and in organic.

This reality, we try not to generalize, deserves a reading from the combination of a nosology of the body that is just the outline of the theory of five elements, which has the power to adjust to the geography and culture of the subject studied and even be enriched by what they have built individuals. But the body has a process according to culture, builds alternative networks when threatened, this interior wonder with a rigor that warrants an interpretation going beyond the objective and subjective. These schemes in the present have been interpreted to refer only to what can be seen in the body, but not what it can produce, built to survive the schemes you want to experience is the paradox that has shown us this new path. Biologists who have studied the process of bacteria and viruses have given us some answers to understand this travel route inside the body[192].

The expectations are huge because we can go to different ways to build feelings and emotions; in this sense, the creative process is known by an accident or personal work. Psychological scientists have inherited only fears we enter the process of a human being with other ways of being and of constructing, hegemony and standardization have been at the point of denial and create thoughts fixed and anachronistic; however, born health workers who have to face these new challenges and are building with their body these new readings. Discovering this new reality in the body is part of an awakening. The most comprehensive is given at the encounter with yourself to know who we are. The emotions and the organs are the elements that combined and put to work can give us new expressions in any field you want to explore, even going beyond what we can now think about the social world built, only that the processes of appropriation and domain should be changed and given new ways of coexistence between subjects.

It is worth turning to see the people who have lived with systems that have more than five hundred years of existence, something to teach us, Why them? Because modern societies were built on another principle;

---

[192] Ilya Prigogine, ¿Tan sólo una ilusión? Una exploración del caos al orden, Barcelona. Tusquets.1983.

which is a principle of death for the body and emotions. This means that the effort is not so difficult, because there are people who have traveled the road without this logic of the body noted. With these reflections we consider it appropriate to continue the investigation of chronic degenerative diseases.

12

How can an emotion be installed in a body?

How does an emotion settle in an organ? The question should go further and also cover how is the emotion received and what is the installation process in the body.

The receiving process occurs in the stomach, as this is the organ that receives food and any emotion or impression, whether exogenous or endogenous. That's where they stay and give effect to the following steps, and the body can distribute them to other organs. There are many factors that come together in the moment when an individual leaves develop emotion in the body, and the mechanisms that move within the individual, along with family memories and body are important in understanding how to set this process in the body. The above is a view that attempts to articulate the complexity of life, which cannot be locked into an absolute paradigm.

The body aches are an indicator, something it is malfunctioning; these signs are important part of a process that is not listening, but sometimes can decrypt the messages and thus prevent bodily deterioration. However, this is something that goes beyond those constructions that society and its test suites have been assigned to the body in a world that is not theirs. Perhaps we should also say that people are not what they appear, they have built images of them in a society that puts them in this dynamic and compete without regard the body shape of their lives; so, tachycardia is only the expression of a world full of emotions that are not processed, and this leads to the neatness of a moment that explodes in a body and atrophy.

The interconnectedness of the body is an expression of society and its values, culture, morality, who are embodied. Emotions are entities that are allowed to come to life: the installation of a "police" in the organs; this can be seen in the process of cultural and religious history of a population, where elections of the subject allow him get to the bottom of a sort of fictional reality on the outside and keep the appearance in a society full of masks.

An emotion can do what a bullet or stab wounds, even go finer points to end the life, with the body; are forms of expression of a mechanism that can destroy or stop life.

If an emotion enters the body through the stomach, we must understand this process of inclusion in the body by way of the relationship that the subject establishes with ways to manage his emotions; an individual who faces circumstances can not leave words without a meaning, makes the relationship with the modes of feeling and understanding the process of its truth and its link with the emotion; in this condition is the subject that should make a relationship with emotions. To receive and file somewhere on his body, the range of emotions facing the set of meanings in the body, this means, significance is embodied and causes an individual to possible attempts to decipher what he feels and that it the world of meanings shelter is just how it feels. We can try to go deeper into the internal processes and what moves there when an individual begins to feel hate or have mixed emotions with his inner peace.

We must also learn to decipher what individuals do with their body and open a way to find those emotions that are woven with hatred or love of the human condition; an emotion not related to these processes do not seem to make much sense. What matters is what is not expressed, but what we do. It is feasible to say that emotions are progressing in social life; passions, desires, have installed their dominance in society and the body, that means that the body is at the mercy of emotions and is the habitat where they feed or constructed to express or move the body to destroy which is next door. Thus, the meanings and representations can not escape this new element is embodied in society and in human bodies.

To understand this process requires an epistemology that is not crossed by reason. We have a body, but it is inhabited by three bodies at once: the brain as an autonomous entity, the body as a place to live and grow emotions and bodies, and emotions that inhabit it and have taken their own life and do not let the harmonious relationship is established.

This involves understanding the processes of emotion and autonomy to the body. A body is torn between doing and not, stop or speak properly is part of the intricate struggle that takes place inside, so, representations and meanings play an important role because they are diverting the construction process of life, resulting in relationships that do not include a passion for life and only incorporate the feeling and desire, the passion to feel, yes, feel becomes the central point of life of individuals, whatever you feel like synonymous with being alive.

This should alert us about the importance of learning to bring the emotions to a level of interpretation from which do not affect the bodily life.

If emotions have become a new *body that inhabits the body*, is likely to be developing new ways to exist in the body itself; this implies that emotion can be moved to other planes of the body. For example, ways of presenting symptoms is something we can go through the same, how they develop in and by the individual; an emotion in the body has a range and can continue to eat or get a space in the brain, which implies that it must become a new physical reality. This corporal new reality is not for all individuals, but for the industry that has evolved with reason, this factor is very important to establish the levels of construction in the body by emotion. We can even think about the ways in which emotion is expressed in the body, because sometimes the brain needs to experience the effects of a relationship where there is given the nutrients from the bodies, which implies an internal dispute. This implies in turn learn to see the body in another dimension or relationship between the organs, see the body of another way is important to establish the levels of construction that individuals make. A body may have a gap between other organs and can even make life becomes an instant, and thus may make humans give new variations on the types of studies that give the body to establish a diagnosis or type of treatment: understanding that the rate of glucose does not drop even with insulin injection, how to understand that pain does not stop even when fed a lot of anti-inflammatory. The body is in space of an internal struggle to establish territories, and psychosomatic disorders are no exception, as they have become a sort of public demonstration of an internal conflict.

Deciphering these messages involves not only accept human being as the space of expression of a social conflict, but also understand the hope of a subjective world that wants to escape from a too complex reality for rational thought that makes its home in the body; this complexity of reason and stay in the body only gives the body more difficult in an effort to live by the complexity of a world that can not be hidden in the flesh.

Human beings have divided into fragments that inhabit social spaces such as work, home, school or recreation spaces; we are a particle moving in the space of society, which has built a new plane to establish the levels of cooperation, but the body does not succeed alone, and faced with climate change, the deterioration of soil, water and air, but still refuses to establish new rituals to perish. Given this, we must build options for a body that is without will, without energy balance, no harmony in it, that's what we do today, but where to start?

Perhaps we should first understand the internal process of the organs and emotions. In modern societies have become more complex mechanisms to establish new processes, and that requires us to open a border that seemed banned for no reason; the crisis of reason is a first space in which the problems have been aired for a reason that refuses to die in a body, it seems to have wanted to do his own will: a reason has become unreason.

We are the only beings that have cultivated the extreme right of subjecting the body and become fragmented units without a junction; two bodies in one may be the result of a culture of reason, especially a construction that has not become alternative life. Make that emotions are made of a life related desire is only a possibility, the other would be to convert the excitement in a mechanism that establishes what matters in morality and ethics of a society, and that subjects the body because the reason link the appropriation to establish mechanisms to unlock the individual from feelings and embodied in the body.

A child is a body that is crossed by this circumstance. This means that the process is not free of construction of sensations and emotions, on the understanding that forms part of a family process to set the modes of feeling or claim in their daily lives, this process becomes a reason embodied in or create a style, if you will, a personality, which undoubtedly can be changed by social circumstances, personal or by the exercise of personal choice.

Use the elements of reason is not the way to our case is only the recognition of a reading that prevents the other possibilities of the body at the meanings and representations that come to life on their own.

That's what interests us, the construction process may well have two paths, one processing and one of statism, which are two mechanisms to cause the body to be given their own rhythm and with it the tone for the new relationship body-symbols, or internal processes that establish the new frontier of search of facilities in the social norms and rules; which is why morality can be overtaken by the new dream of a world that wants to go beyond the need of life a new look for ways of living. In the logic of breaks is necessary to address the epistemology and the value of new construction. We are what we think or what we have said we are. Perhaps a quite faithful to this principle is youth, which is a denial or a kind of hope for the wishes. The world of youth has become a hope to move as much pressure on what it means to live, and it is possible to have a miserable world to become the hope of a better one.

Thus, the body cannot be read with absolute or deterministic view, should be as ductile as reality, and not stagnate in the development of internal or external processes. The world of emotions and feelings is conjugated and gives life to a body that suffers from not having, the suffering body is in the process, it's just a great moment and hope it is not to defeat a thought becomes diluted and doing to the *establishment*. Here the point with emotional and sentimental processes people live, and thereby lets out an option for the construction of new possibilities for a body that approximates the alternative inside, especially on feelings and emotions. This means to consider the ways to go into the body, which is sometimes only through desk studies and not by the construction process of the individual and the various ways of feeling may be extended and given a new approach or a new sense of the importance of doing things in the world of feelings.

If we assume that the body is a possibility to grow with emotions that do not dominate it, then we must accept that emotions can have their share space with bodies without getting them to work outside their immediate area, which are those bodies. The concern comes about because in recent years health problems have increased in this organ-emotion relationship and psychosomatic processes have been difficult to unlock and unblock, promoting the bodies deteriorate without the possibility of doing something with them from childhood emotions. Culture is not only a principle to which bodies are made of a shape, so is the language and its representations, a human being in every sense of the word is something that is at the edges of established culture, which means he can develop more and thus give a little of what occurs inside the body, this means, a human being is what may develop in culture, in its time, but needs interior work to make given step in the construction of ways to be thinner and complex. Now, in these times, we go a little to the bottom of these reflections on the body and its benefits, perhaps if we are more demanding of ourselves we can make a person become another not only with the attitude, but a long process frame includes a world that does not correspond to those that have been established.

In short, the body is the space to overcome the organic and in harmony with the emotions, the problem is that interpretation has become a stranger to the pleasures of a cause does not give answers. We are the body that we have imposed, but also the emotions, the emotion-body connection can become established and cause the brain to make more complex emotions in the sense of having autonomy, making it possible to give a different body. An emotion can also give life to the body, going through the blood or air can be used to clean the same body or cause to be unlocked, without being installed in memory of the body or to prevent it being free. I think this because they should seek interpretation to the problem of the psychosomatic and thus make the body more healthy and longevity, the quality of life to rise and, thus, can increase the life of new expressions in the body and society.

The body is a space that can gestate new emotions, new feelings and, in particular, preserve health, is a great choice on theoretical models of the past three hundred years. This means that we should give a reading and a method that does not break into fragmentation. This implies a confrontation with what is established and made in the body in a society that refuses to open new processes to the human condition.

For us the body is the realization of a complex process when doing a mix of ideology and the hopes of a world that wants to control and carry the culture of fear; fear as a tool that denies other processes of the human condition and even establishing a mechanism that can not do the same for everyone. What has happened to the bodies in these times not interpreted with other possibilities, in contrast, is full of masks and laziness on the work in the body, that means you should not make a world apart from the established, but to break with the established to give new interpretations of life. A body subjected to the pressures and values of a culture can create something very compact that does not let the new sensations and readings are understood as options for viewing and live in this world differently. The known theories have become carried banners and assume that a time is the cross we all, regardless of social and personal processes.

13

The gift of speech as life therapy.

All therapies must go to the body

All psychological therapies that circulate in our environment focus on the brain or emotional states. But one thing is certain: do not go to the body, because it conceptualized as a nuisance, not as the space where you can build; and shall seek the emotions and sometimes the ways of destroying them, which, of course, is a futile attempt considering that emotions are inherent to the organs composing the body. This is the point that interests us, the verbalization is the expression of a type of rationality that want to rid the body and make the symbolic representation of reality that departs from the body processes; hence the value given to the mental structures, linguistic structure, the meanings of a word becomes sound without relation to the body. The metaphor is illustrative of the Buddhists say the head has feet and in modern times has shoes. So the head and body alone walk a different path. It is said when people think a lot. Even we can say that people spend 80% of their life thinking and 20% sleeping.

Attempts to forget or deny the body only takes us away from a fundamental principle: the dialogue of the body. This means we are talking about a rupture with the reason for using the word. Thus, the word becomes a free entity of relationships within the body. If we recall the principle that man is a verb, then the stronger the case that the body is a nuisance; must be remembered that religious devotees flagellate trying to forget or deny the passions and desires of flesh.

This proposal of reason, rationalized, has disrupted the principle of harmonious relationship between man-nature-body, which includes just the dialogue with the passions and desires of the flesh. I start from the idea of conceptualizing the body as a microcosm that is part of a wider process in which the planet is represented; this means, the body is a small planet in terms of composition, it can find analogies with various rivers, seas, sky, earth, day of the year, the seven emotions, the fire of the earth, metals, water, wood, seasons, body memory, etc.; but most importantly in our case is the relationship with the interior of the body: organs and emotions, the liver; to the gallbladder rightful wrath; the heart and small intestine, the joy; the spleen, pancreas and stomach, anxiety, distress; lung and large intestine are doing melancholy; the sadness, the kidneys and bladder, fear. This relationship, which is characteristic of human beings in their management, can become part of a movement that breaks the internal relationship of cooperation that prevails when any emotion or organ within the person. The balance is being built between the bodies and emotions are expressed in many ways, especially in a balanced life, where an emotion does not predominante every day, weeks or months with chronic degenerative effects in the body, which has been applied. So we give more arguments on the case, but the interesting thing, and I want to share with you is the voice, the word that springs from the body, as we express the emotional states of people and, therefore, the internal state of the bodies in terms of disease or imbalance of energy. To be more explicit about the word, the prevalence of an organ is reflected in a style of speaking, the same happens when an emotion is dominant. People's word has a tone, a tone that indicates your emotional state, which is also associated with a taste for certain flavors; if you feel melancholy or sadness you prefer the flavor spicy; salty if fear; if distress or anxiety, the sweet is what is demanded; and if it is sour, anger prevails; it will be bitter for joy. We loose nothing in this concept of the body; the idea of fragmentation does not fit into this logic where the word is not as free as we are led to believe; is not the verb that gets into the body and is making nest and takes over a poor body who does not know what to do with the verb, although the myth says that the man was made of word and reason, therefore denies the space which makes sense: the body.

In this sense we can understand that the word is part of the outer and inner life of the body and can become life therapy, in resource for learning to live listening to the inner movement of the body. Margulis put it well in his research on microorganisms: the origin of life is the principle of cooperation, not competition; demonstrated that cooperation has been possible to build a body in which the organs work together to keep alive a body governed by the brain; in our time has become the tyrant who does work for certain organs from codependency, a desire to exalt an addiction or an emotion and the action is invariably the word, to make verbal or physical subterfuges to route the achievement of a goal.

In the case of a balance between emotions and the organs, the brain does not exceed their demands, whether glucose, opiates (external drug) or the endorphins produced by the body; then we hear the word measured, without noise, squeaks, free of complaints; the person may hear, tolerate and even behave in social and personal life with a sense of cooperation. Of course, this idea breaks down and crashes into the circumstances of our time, that encourage competition and make the word carries the prevalence and frightens the listener. Shouting is to impose the significance of the emotional state that denotes words that relate to an organ, although the use is justified on the principle of a "dialogue" with hands and eyes, which are the expression of a resource made of rationality, the intent to convert the body into a resource that is not articulated with speech, which builds the network of international cooperation of the organs. How to help those who want to break up, if they are part of a microcosm; how to deny it if needed so that speech is the expression of a body having a universal language and universal principles: every body eats and defecates. Everyone wants, everyone suffers; in this small detail is the difference between those who want much and those who are repressed. Their bodies have different feelings and expressions, they can say they are original, but not, as in the end the human body is a complex construction of culture, feelings and emotions that are related to the family and their ancestors, food and flavors with a taste for certain odors and emission of the same, our PH *Pinche Histeria* (does not mean fucking hysteria), even our dreams and nightmares are not exempt from this socio-historical-cultural process. Thus, the construction of the microcosm called body has the use of the word and appeal to it to balance a concept is not new, but the difference is the conception of the word and its relation to organ and emotion, conceptualize the word as a possibility of understanding the body based on the principle of organ-emotion relationship, which is where generates life, then we can say that the word is a possibility of life therapy. (You can even see with patients of certain organs, if the lung, large intestine, voice, words will sound sad; if the kidney or bladder, his words are full of fear, if the liver and gallbladder, he is angry with life and his words will have complaints and grievances; if heart and small intestine will euphoria and joy in her words will be optimistic. In the case of the spleen-pancreas and stomach permeate the word is anxiety. Clearly In many cases the combination is more than two dominant emotions in the subject, even three or four, talking about a process of disintegration of the microcosm. It is common for some patients remain in a tone of voice (or the word hurt if I use the term).

We are accepting the existence of a microcosm, but some people are bent on denying it, the fragmented vision wants to simplify the body physiology and reason, it is better for the social relationship of competition that promotes individualism, atomization.

The dialogues of the body are not different among people, are so common that few will be presented today, another tomorrow, but come in an expression worth seeing when someone says this does not happen to me. The internal dialogue can go hand in hand with the bodies and give us another language other than verbal; for example, the internal motions is a process that is related to the external motion and do not speak of dualism, but the expression of continuity what you have inside and external continuity. However, the society of our time, we propose the use of a construction which gives meaning to a personality who spoke by cell phone and becomes the person who just leaves messages for avoiding dialogue in front, so the body dialogue is interrupted, rationalize all you want, the shelters are contradictory, do what you don't want to do.

But talking to the body and leave a message hidden in it only shows that most people are feeling that something is moving and wants to link with the outside. Where to see if interior peace does not exist, it becomes not easy to build in the body; it is always postponed for another day, for new public relations. Such people become greedy and unstable in individuals of dead word or fat finger for pressing key buttons.

The body and its word is not something that the psychologies are considering, because it demands another philosophy, another reading of the messages inside the body, the movement of bodies is changing and it is difficult to read for those who are hedonists, who only want to be listened, even though they talked five minutes ago. There is so much disease and dependence! Here the word is used for social survival and usually does not purchase commitments; retracts easily because his body is a permanent imbalance; no matter that today dominates the liver-anger and at morning or at the same time, dominates the kidney-fear and at night the melancholy and sadness are expressed in a cough. The word is not the principle of balance in his life. Empty that pervades them is a personal path loss, and usually has words to express the emptiness you feel in your abdomen, which is the corporal space where emotions are received. Decipher is an art that poets and storytellers exercise and can give us words to express what we feel. But it turns out we cannot put words to all of our emotions and feelings, but it is a fact that when we tried we feel better.

The speech has become the tool that links us with the message body; exacerbated the word as a symbol that gives meaning to the bodies of others and themselves. It is clear that an organ is no stranger to the condition of speech, but involves knowing the routes of expression to make sense of what I say; I try to explain the matter. If we start from the relation of the theory of five elements, we can understand the starting point in the relationship body-emotion, and that does give us a proper sense of the human condition in a deeper perspective on the relationship of speech or the word and body.

We have a body that is not only the verb, as I said before, it is also building a relationship with the inside of organs and these in a building of emotions and anger so he goes to the liver can be expressed not only in the face but also with words that hurt, tear or stored in the body of another, or if we are more illustrative in memory that prompts him to find someone to blame for the reasons given in his words the other aggressive and can not say no, anyone can identify that a word has a purpose. Are the body and reason which are combined in a tone of voice that can be faked or open plane and the recipient may have implications that correspond to the word; which is why I think the body with its organs discuss the importance of the word expressed and if accompanied by a gesture or attitude or even a way to wrinkle the face, either laugh, grimace or frown, the significance and implication of the message can be definitive. But as people spend trying to decipher the messages, then construct hypotheses and move with meanings that do not correspond with what you feel and what you say. Let's see if someone tells us something from the heart. It turns out that the heart is an emotion that brings joy and tell us in a serious tone should doubt, is scheduled to speak from the heart, but the word sounds and suggests not. When we discovered that people speak with reason and not with the organs we can understand that words are empty and do not reach the stomach. Usually felt in the stomach the first feelings of emotion and then can travel to another place, which is why listening can be done in various ways, perhaps we should have the repertoire of what is said with every organ in order to decipher what you are saying or what we want to say and even what is behind the speech, a kind of hermeneutics of everyday life, for language, words.

If anger is the space of the liver and gallbladder, it can be combined with other emotions, like joy of the heart and small intestine ... the point is to know which organ is dominant in humans. Of the five elements wood, fire, earth, metal and water, and ten-visceral organs, must be what is dominant inside the body. Some minimum standards include: knowing which organ dominates dad or mom and then make the connection with the organs and one's place in the family context that we are the expression of an organ that dominates the family and therefore subject to the other or away. Hence the words that express or attitudes that are grown in the space of the family. The word that occurs may be the spearhead of a space that is constructed within the individual and so the words can have a disparate meaning in him. Certainly the tone and matter who says it, the activation of corporal-emotional memory can articulate a response that is expressed by defense or discussion on the matter, a person or two can not be free of an emotion that dominates in the family as is the generality. What do we mean by this? A subject that is the expression of the emotions of the family and therefore the dominant organs will be the expression of his corporal life. The word can dominate any space for expression and therefore may be a necessity to speak of the organs (spleen, pancreas), a lot of anguish, anxiety, always wanting to explain everything to denote uncertainty always a "but" to any idea. This uncertainty makes the person always has something sweet in the mouth or should consume unreasonable amounts of sugar, with the resulting obesity or diabetes. However, talking becomes an inner need that is not quiet and sleep is broken; the complex inner relationship is reversed. This is what I refer to as the new reality of the word body in the society of our time. Hearing a person we can denote their emotional state and therefore its potential illness and imbalance in the body. The relationship emotion-word is the space that does not seem to matter in everyday life. However, we believe it is the spearhead of a new body style of expression, affording another reading to internal and external processes of the same; this unit is so compact that you can not divide or fragment in the language or the organ or even an emotion. I think the emotion and the word is related to the organ and realize that new construction is done in a technocratic society in the use of the word, for example, the telephone as an instrument that can only develop reason and speculation, or talk without seeing, only exploits a level of anxiety in all circumstances. Think or try to interpret something that is done vice or habit and even try to guess develop habits (saying, "you feel this way or something has happened to you? Isn't it?"). We can say more about it but now the interesting thing is the point of this language and the five elements. The voice expressed in an organ cannot be free, is built in that inner space where we must take care to balance the body.

There is, however, another important element in the word and it is the intention that crosses it, its burden of meaning. It moves the body and especially the organs; reason why an illustration does not make any difference with a little concoctions or ground, etc.; which makes the word may have implications for "evil" is the intention that puts the subject, it does harm. So a spell, which is the expression of a meaning, does impact on others as it touches parts of others, otherwise it does not work. The intention is injected to any event can upset the equilibrium of a body and that does damage, if disrupts the functionality of the body, goes to the body and can sicken or kill. But, what is the power of the word, what does the spell? The point is that it does with the organs and sends their energy to hurt the other body, wishing do badly to someone works, but also wishing do well. The words themselves are not bad, what the difference is the intention and it is built inside the body, some people do not know how, but they do handle the energy of organs and channel their desires of any kind, but again, the intention is that moves, that counts. The distance between talk and just wanting only two different times within the body may be the word but also the mediator may be the intention that you define. So I feel it's not normal, in contrast, what the difference is the way how to send the word witness as to irrational use of the word: the ancient Mexican shamans say "I have no face, no voice" or word. They said so to allude to their knowledge had not yet strength to make it work in the others' intentions. So the word for them is very important, whereas in the city is largely discredited or prostituted. The words are worn on people, grow weary and lose their power, then no one believes them. So a leader who loses his voice is better to run it or shift it is then very problematic because no one believes him and begins the policy of the simulation: pretend to understand all that work, who pay them, they speak from the heart, now if you are scheduled to take forward, etc., but the fact is that the word has become the swap. Sometimes someone tells us or lend us a secret pledge, lack of feeling words is something of cities or megacities, people are alone, not talking to their bodies, speak conventionally, with formalisms, with simulation, their laughter and their words are part of a ritual of survival. The solitaire has to take something to sleep or talk to a phone line x; the empty spaces of word will never be filled just by talking or revealing personal misfortunes, the secret passion of the flesh. No, the word plays a vital point in this direction, which is why I say that it is necessary, a kind of hermeneutics of everyday word. Giving a different meaning and is not difficult to learn, just know which organ are we talking to discover our shortcomings. If we move from one organ in the day, it gets difficult; we must go to the rupture of a predominance or expansion of the body it makes it a thrill in it exalt the voice. It is wrong to have a single organ in the word during the entire day; may well be as the ancients Taoists; the day has all four seasons of the year, the seven emotions, day and night. Then must live them as they are, but in our day is that it can sometimes be impossible or difficult to control, to fall prey to the haste or competition.

So far I think the idea of understanding the word of life therapy is associated with a different understanding of the process of talking with the body and especially to understand that emotions play an important role in the uses of the word; link to them a different meaning in the body is part of a new reading of the epistemology of the body that demand our time.

14

The gift of speech and its effects on the body.

The spoken and written word is a dangerous tool also it is the most sublime thing the humans have developed. It has become the principle of living or dying; their effects are lethal when directed with intention, can destroy any relationship or person. They may even let a human being to become the number one at that to make life a ductile instrument, between fantasy and hypocrisy.

Human beings are the bearers of words and their meanings are the literary images that give users an idea or even a new meaning, you put words to feelings and sensations that are produced in the body, the emotions of a time; recover the spirit of an era and we inherit the word games of our generation immediately preceding. He even put words to emotions and turns them into a spear that pierces the conscience of the instituted and through that can feel what is not expressed in words. Curious relationship with the words that express an emotion or a state that can only be put into words the poet and writer.

But when the word escapes from these professionals can be part of an ideological or religious speech and may even become the means for someone to find a cure for his thought or consciousness, the word as a resource to download inner silence is not only has become the means or the use of some professionals who lend their ear to hear the words that have meaning and a relationship with the history of personal life, which can become a soliloquy or the beginning of a private space of the social complaint: the office of the psychologist or psychoanalyst.

The word can also be a good resource for doing business. No vendor or speaker can let me lie; even the preachers live this way. But now we need to go a little beyond this ideology to see the word as a means to control and give meaning to the hustlers of our policy or see it as the resource for building a truth that can be integrated with a deep sense of the simulation, the speech band is not as dangerous by itself, it is when people have a thing called wishes and give them the course, even the sense of the words to make or break the lives of others or staff.

This was used without a warrant and uses what being a man who gives meaning to the words; anybody can prostitute the words by not having them, no relation to spoken and what is done. But the word becomes an ally or an accomplice to evade responsibility for their actions, is the resource for creating smoke screens and it is a new principle of escape as it dispels the lie. Thus, the word can be a pimp in the mouth of the simulators of work and politics.

But the word and its process in the body can be a cause of various forms of construction: the meaning of a word may be rooted in the body of a person and a process that may culminate in the destruction or dismantling of a complex process as is the personal lives so the body and speech in our time have become a basic principle for mental and physical health of individuals. For Harold Bloom, the man who managed to put the emotions in people was William Shakespeare; he gave meaning to an expression in the body and emotions are socialized so far in his theater. The expression and socialization of emotions does not escape the words given to him a deep sense of inner life, but inevitably the action was and is the word. For that reason we can see that the memory of the body is the experience of a culture and history; the construction of emotions is a historical-cultural crossing bodies; but the use of the word as a tool that gives meaning to the man-culture and mediation of representations and meanings do not escape from the idea of being what you want. A human being is not in this articulation of understanding emotions as the messenger inside the body and how it materializes, is made flesh in the corporeality of individuals, are part of the complex microcosm that is the body, have a relationship with organs, with the understanding that emotions have a correspondence with the organs. In this relation, we find that the words expressed within individuals and from there you can make a human being can change or destroy his life: the word as the expression of an internal relation is not free to make an elaborate subterfuge body to defend the truth of the heart and is not expressed by the tongue.

So the words have to reflect the inner state of the body and have to change it in turn: this process can not be understood if we fail to see that emotions have a strong sense of tradition with the bodies and that makes people can not escape its dominance in everyday life, this means, you can not avoid saying someone is pure liver or pure stomach, large intestine or kidney, including heart, and that is expressed with a reading of their words and their tastes for flavors he prefers. The correlation with voice allows us to understand that a person uses words not only have social significance, but also are part of this process of inner organs. The story of his words is not without family history, the meanings and representations of simple or complex words are linked with how to feel them or take their meaning depending on who do they come from. Thus, words do not have the same sense or meaning if they come from known or unknown, relations of people are not the same. The words should be in relation to the body and especially to the organs; hence we can understand why some people talk a lot and others speak little. The action of speaking is correlated with the internal states of the body, so the word has its effects on the body and can make you sick, untie sensations, liquids, juices, endocrine changes, heart rhythms, heart palpitations, headaches, etc., and I'm not speaking in the psychoanalytic sense of the term. My reading starts on the consideration that the body is a microcosm that is built into the relationship between man and society and the process of culture, but an important consideration is the link between the five elements and body organs, for example, liver and gallbladder, correspond to wood and the proper emotion is anger; heart, small intestine and emotion is joy; spleen, pancreas, stomach with earth, and anxiety, thinking much, are proper; lung, large intestine, with metal, melancholy and sadness accompany them, and kidney, bladder, water and fear are related to these two organs. Thus, we find that each organ not only has an element if not an emotion that can cause damage to your body and vice versa. In this process is that the word plays a very important role, because it is the way you express an emotion and a little further, as shown in human relationship. The effects of a speech words that are not exempt from a building process that makes the individual in the way of seeing the world. Significance and representation can become to build new ways of seeing or feeling the emotions in the body, but the value of this process is the word as a tool to transform the individual and organ-emotion relations. Of course I must say the body process is not free of disorders that occur in the geographical area where you live; this means that it is not conceivable that a human being it is independent and free of social processes and premises in which it operates. This involves a complex process of correlations between individuals, for surely a human being is the center of the earth and can not get up as the only people who have a taste for knowing the world. No doubt, speeches or words cannot be left to see according to the places where there are, hence the value of a view that allows us to go beyond the simple representation of an icon. The word becomes the beginning of a story that leaves no room to hope for a better world or a speech where words do not create expectations that a human body is subjected only to the word; the dangers of speech that does nothing but smoke and tells you what to do without going through the reality or fact do, certainly does not have a relationship with your body, has not lived or felt in your body. The detachment from emotions becomes something that is only fun and can not wait to become a new concept as a person to be the words recourse only to the simulation. So the word is not related to the organs and emotional states in danger of being empty, no longer hope to be covered with tracks.

The value of truth is when it becomes perfected, but the word, which simulates, uncovers and cannot be part of a truth. For the body, the word is substantial and substantive to give new meaning to what they say, people living or daily work.

A speech by words without heart may well become a dead letter. The microcosm called body becomes the space where they are built and words can be the place where the movements transform the speaker and the listener and that puts us in line to propose that a human being who is in inner harmony and takes out his construction the complex relationship of the word with the interior life style and it confronts us with the complex ways of making the reading of what someone says something or claiming this or that idea. The value of things is not what is considered to be true or false; is the way individuals live it inside their body, which means that the word is the result of work that goes a little beyond linguistics. It's not just a representation of the world or the ideas of men, is a process that is related to the immediate geography and style of living the body, how to exercise the emotions and, if there is doubt put style momentum to the relationship with the world of representations in intersubjective relations of individuals. The complex relationship that involves building a word not only a work of style or grammar; no, is the complex construction of a body and a spoken or written expression. The body is, of course, not only the waste producer, is also of words and ideas that involve meat or organs and of course the condition of the spirit may well be the extension may well be the continuation of a culture or history that is tied into the meat so as not to let emerge the new or true of a body sinks into the frames of desires. This means that the culture of hope that we have inherited not allow us to be what we want to be in our time. This implies a deep confrontation with our interior to get rid of the oppression of a culture or a historical myth that emasculates what we can be in our time as individuals, free is not easy, for example, to say that Mexicans are lazy and assholes, corrupt and so on. Especially when a Nobel laureate Octavio Paz said, it becomes a sentence in the body of a teenager who is just beginning to want to be him and is this conviction that either makes him feel castrated or sinks into the worthless feeling. The book *El laberinto de la soledad (The Labyrinth of Solitude)* is an example that illustrates what I mean.

Especially if we consider the complex condition of not knowing that words have an effect not only conditioning, but change inside, the way of life becomes the beginning to make sense of the ways to make a lifestyle be coercive, if not predatory within people; this process ties the inside microcosm and can render the idea of castration.

This process is something that is not turned on inside the body, which is why I say the word becomes the instrument or tool to kill or rid the body. The conditionality of an oral history conditions us and we decline the other side, exercising the office of trying to feel with the belly button or the heart. In the Taoist tradition says that we think the belly button and not the brain.

The word and the body have been dissociated into something different in the subject or, if you will, individuality, because today's discussions that have arisen seem to go all the idea of having a body is only a social and history and of course, leaving aside the process of a body that the individual is appropriate and can lead to the release to be him and take what it can contribute to this consumer society. We say that a human being who is different will have to start with the handling of the word and its correlation with reality and its consistency with what he feels and what he shares with a society that every day is diluted at the beginning of wanting to homogenize the meaning of life, hence the body is the only place where we can again be back on top of us, even when the society is in this simulation ideology and politics who thinks only of politicians and not the unborn except win those already born.

I therefore think that a word that works in the body allows the individual to be him and give meaning to action to make life a long way with the body. It is not exclusive talk about emotions and feelings, we need to work with the desires that give a contrast to allow the construction of suffering in the body is suffering or wanting to have more careful of what you have, and even not having what you desired. So the word becomes the principle of restoring the body's intuition when we can say that the talk is felt and lived in the interior, it gives a new course to what we feel, can approach the emotions of others and identify that are theirs, but imposed by society and culture.

The term imposed on the body involves the construction of an internal network that can give an identity to the subjects to build a new possibility within the organ-emotion relationship. This involves the subject's emotional castration attending a long way to learn to see through the eyes of the others can not give an answer to what happens inside. The above consideration is based on an individual not only is the product of a society that have arisen and repeats as a parrot. The sole ownership of their condition puts you at the border to make your work with some conditionality microcosm that becomes the last of the representations to live. Given this reality there is not much defense, because the social becomes conviction in imposing a language and words that are not libertarian; become straitjackets that prevent articulate the language of the interior, so it is not easy for any citizen to build new words. It takes an inside job, to harmonize the heart. The poverty of this word-body relationship in the vast majority of citizens is alarm because it leads us to a point of sadness and deterioration making impossible to learn to feel with the body, with the belly button, not the socially instituted; so I think the word has become a tool that can go against the denial of learning to feel in our time beyond the hedonistic sense, build words that give new meaning to a body or the body to make sense of the words and download it to be an act of building the sensations and emotions in language that gives words to learn to feel these days as arid.

Reason why the concept of microcosm. The body is the space where advances forged a new horizon in this quest to make something of the human being humanized. Appealing the concept of microcosm is substantial in the perspective of building words that arise from inside and give it a new order instituted, it breaks the absurd idea that can not be created in our time or that everything is done. Descartes and Marcel Mauss said the body was a machine or a sophisticated technology; I think the body is a great opportunity to learn to be one in the here and now. The language and its consistency becomes a complex relationship with the other, which can be labeled types of coexistence that can range from the continuation of an order to the construction of new representations. The relationship with the body is identified with the dissatisfaction with what it feels like no one can leave lying to say that a person has a feeling of dissatisfaction with life or with the relationship he has with others. That can let you know that words and expressions are not quite complete to express what they want; sometimes they suffer for it. Thus, we find the missing words to express a desire or a feeling that's why a poet or a writer can say it better than one, but sometimes that does not satisfy because the feeling is different and the words are not correct. Creation is not befitting a man of letters, is human beings who are seeking ways of expression to make sense of their internal construction, but not finding routes may become frustrated and bitter. So the body is not happy, does not enjoy or is not full, there will be dissatisfaction. But the important thing here is how to build a body and gives meaning to the process of making his life in relation to bodies and not only social life, this means, organs, and its network of international cooperation allow us another face, another voice; work is related to the process of establishing a harmonious network among the organs to prevent manifest imbalance. What is produced in any state will support the general statements of others and be feasible to have speakers, so there is no guarantee that it is true, given the large amount of junk ephemeral literature circulating in our environment. At the end of the day is the product of an ephemeral state of the body. Therefore, the relationship with the interior of the body should not escape the connection with the history and culture: the word is the balance point to know that the lifetime becomes human and can open new horizons in the network construction internal and why not open up new codes of the millions that have been opened, the medium is the body and interior work with the organs and emotions. So the search for the words cannot be so fragmented, this relationship should be linked to organs and emotions to really release of ideology and the power play and we can have access to a literary quality that we invite the construction of new sensations and feelings to live in this world.

15

FEAR AND EVERYDAY LIFE

Some authors have said that fear is something inherent in man, which serves to be preserved, which has a regulatory function, even that is inherent in existence; we can add that is part of the human body and its balance is synonymous of good health, this means, it is not end the fear, we try to keep in a state that does not prevent the growth of the individual. Don Quijote said to Sancho Panza: *"El miedo que tienes te hace que ni veas ni oigas a derechas; porque uno de los efectos del miedo es turbar los sentidos"* (The fear thou not see that makes you not hear a right; because one of the effects of fear is to disturb the senses) and has reason to disturb the senses is not the same because the reading of reality and the body is not the same, they make mistakes after mistakes that become a kind of curse that seems never to end. It is when people fear a lifestyle, a status of refuge, a source of dialogue, an excuse for not moving, the fear is to become something that controls the body and will be lost. The question always attacks, anxiety is present in everyday life is when you seek shelter in any addiction, any excuse to do nothing, to invent a disease to justify inaction to bring to a somatization becoming the obvious element for everyone knows that's why the other does not move forward and unleashes pity or sympathy in those around him. In fear that logic is a big monster that attacks the weak and fools, gives the impression that it is an entity that goes flying around dark places and one day attacked without mercy and makes children wet and can not sleep alone. One can say fucking fear, so slouch, why does not attack me I am an adult, and leave the kids alone? Suddenly one day it attacks you by walking with your big mouth and that's where the problem starts, it begins to move the endocrine system, glands secrete, among other substances, proteins that alter the body's functioning and symptomatology is complex. Fear is unleashed in the body, we can say that literally invades the personal and social life and there is no escape, must face. It is said that we have two options: to paralyze and do nothing, stay in that morass of terror with life, we become fragmented, we take refuge in anything to do nothing and so many lives are consumed in the present; true potential dry with fear and you see that and regrets it, but look to take it out of that state, but the other clings to its ghosts, fear castrates him to pursue a life that can be free without the stigma of I can not. A second is to accelerate us and want to escape from that fear turned into persecution, no fortress is so safe, no one gives confidence, and we can even say that this person claims that no one understands him. Fear makes us believe we are original or unique, so do not stop, are giving an overactive fear.

The truth is that in either of these two circumstances fear is not beneficial, the body is subjected to processes that are becoming complex in their internal relationship and form a personality that is not balanced, hence the excesses, vices, addictions, desires, and so on. The paradox is not to be afraid.

So where do we walk? you have to ask, if the fucking fear I have it is on the body, head; even the colloquial expression "I will come right now, I will throw the fear," solves nothing, it just makes to rest colon or bladder, but fear remains in the body.

I remember one of my fears, as a child was that the world was going to end and I thought everything would die and people would have much pain and I was more afraid and that caused me the Atalaya, coming every evening wanting to convince my father to become a Jehovah's Witness. My father, a person without fear of living, just say, "those bastards, that the world will end, it will end for the dead and that's it, you have to be afraid of the living, the dead are dead". Years later I realized that the world will not end just because someone realized that words have great power to fear or suggestible and to build enthusiasm for life, just had to find a balance between emotions and daily life.

How to overcome fear you have to ask. There are no recipes, but the first step is to understand that fear is in the head and have stayed in the body. Learn to see it as a tenant is just uncomfortable with it alter the link, you need to take some actions to be put in place, it is established in the kidney in a balance on the implications of living with him, but without their fearful presence in the whole body, have to leave when a dog wants to bite us or when you have to hand deliver some accounts even when they have to give an important step in life, it could prevent us from not paralyze us or become accelerate, one does not know what happens in what we do, just creates a little or a lot of security. Of course there is the fear of ignorance and that is another kind of fear which is close to the negation of growth, in these cases it is necessary to stick to someone who has lived and lead us and guide us.

Living, then, is not so complicated; we make it difficult to advance its what people say, what will happen, what others think or even know that it is more work and I can not do; is an amorphous mixture of stupidity, mediocrity, insecurity and fear and these become a kind of suckers for those with less fear. So I think you just need to do, with the understanding that nothing serious happens, neither the Americans invade us, or stop the sun rise tomorrow, or steal the flag from the Zocalo and politicians are not going to steal in Mexico and If something similar happens in due course will be what to do, but for now just do what you have to do and people will be healthy and free of fear.

16

ADDICTIONS AND SPIRITUAL GROWTH

The process that one has lived is a state of semi-sleep, peace; love of life and above all to make sense of one's life and others. That feeling is something not to be missed, we must cultivate a strong sense of life, a large tree that is rooted in the hearts of humans and is giving food to the hope of new utopias within the society. The tree metaphor I like. Indicate it as an opportunity to transmit the teaching of *attachments* and *spiritual growth*.

Human beings are the result of their time, their father, their mother, the neighborhood where they were born, school attended and work where they are performing and, beyond that, are the result of the idea of God, the idea of happiness, the idea of pain, suffering and desires inconclusive. You learn to suffer for what you want, so you do not want to drop, because it considers it necessary to cling to something to be or to have an identity, even the culture of complaint becomes a source of identity in a kind of vicious circle that makes us become attached to the complaint.

But attachment is also a possibility to grow. Here are some possibilities. The first is that one must realize that our attachments become a burden to live and a way to awaken the body. How is that? An attachment is only the beginning of the domain of meat and food of rationalist thought, in other words, the stupidity and banality make a mixture that becomes a tool that inhabits the body of citizens and makes jokes heavier. Addictions are hard to control, being close to someone, the skin feel, smell, hear the voice, to give us special treatment, either punishment or caresses are all attachments. They built the body and the right to not feel alone, to have something to feel or through them, be a part of another, which justifies the existence of a way of life; with them is endless the hope you live, is built a sense in feeling and seeing life; and the senses respond to this attitude of taking the world by recording the hedonism to be fine, just feeling and feeling is pain or pleasure.

The body is just the space where the existence of rejoicing citizens who are sleeping with attachments, that road is much hesitation, much claim to life and tell me what had to happen to me, why I have to fall in love, why do I have to make love or be loved, why need the presence of someone to be or feel that I am required or necessary in this world of desire. The need to listen to tell us you love us or want us has a deep sense of attachment to the words of others.

Perhaps it must be because we did not learn to love ourselves, we cannot live with our bodies, and our feelings and we want someone to love us, tell us who loves us. The problem is that the other is the same and calls never say that we love him, because we expect only tell us so, we ignore that the other has its desire and need to be fed well. Here the problem of conflict and denial of inner growth.

Not just enough to know that I have registered the wishes, or be discussing with them like a fight I have to give in honor of the family or my pride that becomes calling card to those around me and can say the Other "How that fellow suffers, but never mind, he has to learn." The truth is that not learn, simply suffering and more attachment to their pain, because that gives the opportunity for others to look for: someone cares. But really do not mind, cannot be released from an addiction to words. Action is required with the body. Move your body memory.

In those circumstance attachment is difficult to grow spiritually. It may do laundry work of guilt, it will stop being addicted to pleasure, forgive others, yourself and work hard to outside, but the small detail that needs to work with your body, letting the wishes and attachments do not govern personal life to awaken the other inside that can be converted in governing the hope of being one with the body. What does this mean? It involves what I have to bring in and that is the spirit that governs the organs and not thinking, which unbalances the inside when you think compulsively victim of an addiction, this suffering becomes a share of punishment that does not leaves us to see where is standing or where it goes. You make the mistake of comparing with others, with those who are more screwed or which are more developed. The result is that one loses and takes refuge in the claim or flogging, disqualification and learns to see only the negative side of others, which does not bode well to say that I am growing spiritually is just an illusion or a refuge for not growing and not assume that the commitment is not only with others but with oneself.

Accepting to have a body that should not rule alone, I can make him face the reason and the attachments, I can place them in a body I do not love the passion of my desires. This path can be a good place to start to grow inside. Incidentally, what I work in me is what I give to others; otherwise just give them empty words that wear out over time. Yes, they know that words wear, more than one know that the statement "I love you" is not equal to the heat of conquest that two years later, when they are confronted or attacked actors so idyllic relationship.

The consistency of the words is invaluable when a person tells us whose body is not subject to the desires, attachments, have been living with them and give them instead. Then his voice can be heard inside and that is a great encouragement to make sense of someone who wants to wake up in your body, without denying the lived, without accusing anyone of sense, seeing faces of evil, he just wants to learn from what is experienced and correct in a way that is not finished. When someone does that, we can say that is on the way of spiritual development and I do not mean to be St. Francis of Assisi, I think we should be characters of our time, with precise guidance on development within the body, we must assume that we here is where we build and share with others. As Buddhism says, take the path of enlightenment.

Overcome mental and corporal laziness has to do with attachments; there is also an addiction to laziness, weakness, which may have different origins, but we must face if we want growth and spiritual development is grown in the only space we have to do, which is the body.

At the beginning talking about the metaphor of the tree and say that I love, for the simple reason that it is the only living thing that lasts longer on planet Earth. There are reports of some 700-year-old and say, "there's a reason." "What is it?", I'm going to ask and I have to observe how they live to find the secret of long life and find it has no attachments and knows cooperate. Of course I'm talking about a tree that has developed a deep sense of cooperation and not a pepper tree or a eucalyptus cooperative are envious and don't let other trees to grow, do not let others make nests, do not absorb rainwater in a short time, among other silly things they do, watch them die soon. Even the winds toss them, indicating they are not as balanced and its existence is not very straight. In contrast, information about the average life of other trees is 500-700 years. Of course, other trees will not do that kind of nonsense: they are lush, with many nests, others grow around its trunk, fuel and other living beings are in harmony with the winds and rains, its branches grow to maintain a balance and grow in righteousness, its strength is the middle path, that is, share it and may make sense for others to find the way of existence without attachments and that to me is a lesson to be recovered from a being who has without competing long life and let life elapses. I think the strength of the interior is a great principle to achieve balance with a life that flows with the responsibility to make a decent life on this planet. A living as a tree synthesis for me where we can find what they talk about living with addictions and building inside without compromising the identity of being me and do not the wishes of others.

17

THE HUMAN BODY AND ADDICTIONS

2.500 years ago was born in India in the region of the Sakya, a man who discovered the source of human suffering. Siddhartha Gautama is called, known today as the Historical Buddha. Only facing himself was enough to discover the source of pain and suffering of the people is the desire, attachment: people suffer to have, for not to having, for caring, for wanting more. The idea that happiness is given by the goods, money, love, food, sex, illusion, fantasy, rational knowledge, etc., only causes suffering and desire in people, and pain because they never get full satisfaction.

An illustrative example is when teenagers tell their boyfriend or girlfriend; *tell me you love me I will never leave me?* Etcetera. Although the answers are affirmative always want to hear more. Insecurity is part of this ritual of asking beautiful words to give security to keep the other. It can be a romantic attachment or corny, but it is. And it can lead to a co-dependency, jealousy or homicide.

We are the only animals that we complain about everything, this culture of complaint is a kind of obstacle for not walking, not to have options and do not always justify the things in our lives or this life.

You can see anyone and it always has a complaint. Because it rains, it's hot, it's cold or it is windy. It seems they have not made the world to suit our vanity or our ego. Surely no one will because everyone is thinking about something banal that nobody deserves. And this is a first attachment in people; the image of his cosmetology chameleon is an attachment to the illusion of being better disqualifying the world of his birth, when it should have gratitude for this noble earth, which never complains about the silly things that make human animals. But no, this attachment to the idea of being original in a world where nothing is as original as life, not an object to build, is a possibility to grow, cooperate with others or be proactive. This principle of adherence wanting to be one without including the other is the result of a competition culture, trying to be the best no matter who is trampled.

The man who discovered it said that it was ignorance, an ignorance that prevents us from seeing that I have a body that can be a mean to enlightenment, inner peace. The counterpart of this idea is the reason, the rationalist thought you did a great bump to the human brain and forgot body like the harmonious possibility of building a being that may not be delinked from nature, of its nature. Although he has built culture and developed the reason, it is a possibility for living with nature in a rational use of it. This attachment to reason creates illusions, fantasies and false expectations about the business of living. The Buddhists say that thought is a brain disease. Just look at those who think a lot: some think they spend 80% of his life and 20% sleeping. Result: a life of anxiety, hemorrhoids, headaches, migraines, gastritis, colitis, constipation, impotence, amenorrhea or dysmenorrhea, among other corporal goodies. Not to mention the emotional abandonment, flogging to get attention and unconscious suicide. All due failing to keep the balance in life, excesses, and absences. Are not recommended any of them. Always seek the middle of things; the balance between living and attachment to the ideas or sensations only gives guidelines for building illusions without a job. This means that the work I do allows me to maintain a balance between eating and living. And I can be me or have the option to break free of the culture imposed on me by family or institutions. This is a chance to be free of attachments.

The way the Buddha taught his meditation, which allows a revolution in the interior, changing the body to be competitive to cooperative involves a change of attitude. Not to be contemplative, means acting in the geography where I was born, in the culture that gave me a language and intern always do things well and of course conclude, that allows freedom from fanciful idea that one day I'll do this or -so. No, we have to say I'm doing that. That's being in reality, the here and now.

Life has become confusion between the haves and possesses no measure whatever. Although this generates pain and have course more attachment. The result is unhappiness, dissatisfaction with what is done, and the disqualification of anyone who has something I do not have; because of course they are all fools or idiots, unless you say so. There is everything in it to stick to ideas or make others enter destructive dynamics. That way of life, is how one can find that humans have many attachments and you do not have realized, they think are unique no matter how mediocre they are. That is the quality of rationalism, inflate the ego and thus lose touch with reality and with yourself. No doubt this person is asleep, would say the man who spoke before. Asleep because he has not awakened in his body and do not even remember being on this planet, with plenty of opportunities to live and not see that others are he and he others in this dialectic of ignorance of the business of living. So it is easy to find the ones with most addictions in their life, you see them in need of always having something in their hands or head. We explain: those who always talk to hear the voice of so and, those who ask to give them unconditional love, those who never give anything in return, claiming a call, which require time for them, as if life was a commercial transaction. So do not let life flow, to go without expecting anything, just do it to live in the here and now.

Life is always waiting form of attachment; if I am so I get both. If you do not expect any attachments may tell us we are "brain surgery". When one clings to an idea and fight for it is different. The Buddha fought for six years to reach enlightenment under a ceiba and left a path as teaching.

Living is the combination of styles to see the world, to combine the teachings of parents and the wishes of the authors you have read or the others, attachment to things is only a moment in which one can say that the influence of seeing the world is a long wait to say that nothing is forever and everything is ephemeral, nothing is forever. Human beings always have things to finish and that causes them to suffer thinking that something must be your true self, which is an illusion, since there is defined to be searched for happiness. That attachment of a long search to academics, a truth that is endless and goes to the merits of the idea of being someone, as if having or possesing give meaning to life. So many human beings have been covered with expectations to deal with themselves, unaware that they can change their lives starting with the question of who I am. Answered this question may hold the key to doing an obsessive search outside, in reason, in the exaltation of the senses and lose yourself in the enjoyment. The pleasure is an incitement to lose the way to answer the question of who I am. The body is a possibility to deal with attachments and desires, the rest is just a fight breaks out between thinking and doing. I think that should face the body and desires, addictions that we cannot see the reality in which one lives and less have happiness.

18

THE HUMAN BODY AND HEALTH

In recent years, approximately thirty-five, the health sciences in the West have entered a phase in which the forms of healing and explain the processes of health and disease have found a border that does not seem to be able to transpose. The biomedical model supported by molecular biology has had to backtrack on the absolutist explanations. It has opened the door of doubt in the perspective of the human. The body concept of the model, consisting of blood tissue, bone and muscle, led him to seek the physiological process of any court for a possible cure for the altered homeostatic cycle and if you want it simplified as the unit biopsychosocial kind of transplant without cell compatibility, that has not solved the problem well drive for holistic treatments[193].

Note that the above is based on a conception of the human body like a machine that needs change and spare parts and establishing a market where anything that can cure or heal the human body becomes good. No

---

[193] I believe that today the problem of health is one of the most profitable businesses that are known. You can inquire about it at: Fritjof Capra, *El punto crucial*, Ed. Estaciones, Argentina, 1998; y Larry Dossey, *Medicina, tiempo y espacio*, Ed. Kairós, Barcelona, 1989. In both texts it is an excellent exposition of the limits of biomedical models and their link to the Newtonian model in the bodily conception at the same time as show the limits they face today.

doubt the human body is the greatest source of wealth that is known on the planet, large fortunes have been thinking about what he gets, what it is spread on it, everything that occurs and the body is made feasible to buy or sell, marketing has taken over the bodies and gray modern cities. In this logic the man is not master of his life, his health in the workplace the body is exposed to occupational diseases, to the upheaval of a life full of tensions and competition for survival. Remember that humans are the only animal that kills for pleasure rather than necessity.

The human body is then the possibility that men have and make a social life, to confront the limits of their strengths and develop their disease, which has become normal among the inhabitants of megacities, because they share habits, rites and pathologies, democracy is installed via the ways of living and dying. Today, any unemployed person can aspire to a heart attack, perhaps I should say that a respiratory arrest.

There is a widespread belief that there are people who care for our body, it seems that we are not responsible to take care of ourselves. In this idea we find the origin of the body drop, since individuals do not deal with their lives, they serve when they are in the critical phase when the body protests when it can not respond well in daily activities. That is when you think you have a body and when they start in the best care: the changing diets allowed some excesses and begin to see body changes. This observation makes the subjects begin to have a different perception of the moving body, feelings and pay attention to their expectations of life, forms of life can change. The body can only perceive or live in the moment, also gives us the ability to leap borders to wage a compulsive consumer society imposed upon us, which mainly has told us that we must consume bodies, whether they be men or women. But perhaps not the pleasure has become a possibility; on the contrary, it has been stereotyped in the genitals.

The body is the starting point for all industries, pharmaceutical, automotive and cleaning products, etc., which means that the human body as a concept that allowed the possibility of having a better life on planet Earth has been forgotten in big cities. I say this as a way of life, because fortunately there are people concerned about their inner life and doing various practices and ways to cultivate the body in its interior and exterior, which results in heterogeneity of opportunity exists in our day and perhaps fit the claim that is precisely due to the crisis we have passed health official models, models that have a high rate of deaths and we refer to the health system that dominates in hospitals, white elephants. Some authors believe "that when long strikes in hospitals, the mortality of the population decreases, perhaps because they forget the doctors, but know it is because the excessive medication[194]". Allow the body to health workers means that we give our confidence about possible care they deem right and maybe there is where much of the iatrogenic, as a body in the hands of the other only allows us to have confidence and faith that the other is doing correctly. What shall we do then?, Perhaps readers may wonder. Just ignorance of body leads to the search for measures that consider the responsibility of others, when we are who feel and suffer the problems with our bodies. This claim faces a myriad of thoughts, but every day we see the implications in regard to living and dying, the importance of having the right to a dignified death, not one where those affected are those who remain, for as Elias Canetti said the dead man's strength is greater when we see him die, because that is the last of the images, and this burden can live for long. The human body is the point of departure and returns and in turn may be a symbol that you can build better life metaphors though the body is battered.

Thus, we find that the conception we have of the human body is related to how to look for causes, or is it mono or policausal are exercises of fragmentation on life, vision is certainly not far from the perspective of the universe and the world explained by Newtonian physics, where the absolute and deny other possibilities of understanding and treatment of human beings, that is, other epistemologies, other prospects to read the living document that is the human body. The ways of knowing the body to heal varies from one culture to another and so we have two perspectives that run counter to the understanding of humanity and the universe.

We can point to the West and the East as two models of the human understanding and practice, other than how to care for and keep alive the human being. No doubt these two epistemological perspectives have no common ground from different conceptions of the origin of the universe and therefore the human being.

---

[194] See text: *Nueva conciencia, plenitud personal y equilibrio planetario para el siglo XXI*, Integral, España, s/a. P. 63.

For Chinese thinking man's origin can not be understood without understanding the motion of the Earth and Moon, the Cosmos, which is the origin of things on Earth, for the Chinese unit is a basic principle for understanding the simple and complex processes of human being. The unity of yin and yang is in all things of the earth and of course in the individual, which constitutes a kind of microcosm, which works in harmony with its constituent parts. This unit, which gives rise to the vital energy or *ki* is kept in balance and harmony with nature, other individuals, feeding, breathing, upon presentation of the imbalance of these two forces-yin, yang, we present a symptomatology is directly related to the biological and emotional human beings. There is fragmentation in the organic and emotional life of the subject, this means, duality and coexistence are a working principle and diagnosis in the body, not just rhetoric. This principle of physical reality can be violent in the eyes of a thought seeking causality fragmentation.

There are two logics of health, two epistemologies that they can coexist in the same spaces and help each other to prevent and cure. The attitude of many health professionals has been changing and we can find it on allopathic doctors who make proper use of Chinese medicine, especially acupuncture. It takes experience to conceptualize human beings as such and not as a dollar sign, that has led to such a level of dehumanization, which is well worth making an assessment of the successes of a health system that focuses on supra-specialization and only studies parts of a set that can not be separated in practice. Conceptually maybe yes, but not in the service. I think in this century to witness the opportunity to make changes to the coexistence, despite the huge economic interests at play.

19

Understanding diabetes in a unit body-emotion

The principle of non-fragmentation bodies and emotions

This paper aims to show the relations of a body building process of diabetes in the body of women who were treated at the *Centro de Estudios y Atención Psicológica, A. C. (CEAPAC)* (Center for Psychological Studies and Attention, A. C.) with the proposal of acupuncture, which only take the theory of five elements to explain and analyze the relationship that integrates elements such as food, emotions and predominate flavors in conjugation with the intersubjective relations and their links with chronic diseases and more frequent psychosomatic.

The concept of body-emotion is an approach that maintains that there is one without the other, are an inseparable unity that we can see in the daily exercise of existence: a stomach ache can be related to a high level of anxiety, a fear or apprehension, and the frequent excretion of urine, or failing that, with no filter. All sad or melancholy people are submitted respiratory problems, whether high or low, which means that a relationship of emotion to the organ and its expression in the body make a set that we cannot divide. In this logic, the human body should not be fragmented for study in any conditions. However, not enough to say that the health problem is a psychosomatic problem, it is necessary to describe the process that integrates, explain the travel route of an emotion and how it's installed in an area or organ and, of course, why in a body and not another.

Cannot understand this principle of the path of a disease without knowing the parts of the process of a chronic degenerative disease. The first question is how to build it? Starting from where we understand this new complexity that the body has built from its own organization. Is it the body developing these processes or is another factor that produces them? To answer this question we must start from the body itself, that is our reality to understand the organization that builds a body in space and time. We accept that the body faces the challenge of a process that threatens from the outside and that allows the construction of different answers to essentially biological level, which means it is part of a whole complex and that what happened in the past is in the memory and projected future.

Then, too, can say it is a place where you save the processes that forms it and participates actively or passively. With that physical reality we will work. There are pathways and processes, without forgetting that is part of that social complexity that occurs in geography. Only thus can we understand that force generated within the body and is able to lengthen or shorten the life of the subject building mechanisms of the body as a self-regulating system: first preserves life, then defend it and, at worst scenarios, destroys a chronic degenerative disease.

However this does not answer our questions, as we cannot say that's what psychosomatic or chronic-degenerative disease. What does joining the construction process? It can be the concatenation of the body itself and the regulation of biochemical metabolism, we can even talk about the homeostatic cycle, but other questions assails us: how can a body that is capable of forming itself can be overcome, spasmed, contractured or may produce a respiratory or cardiac arrest and can stop any functioning of the organ? We cannot respond to this new question without articulating the logic of biology of the body that assumes a certain level of perfection, a condition that is just what allows us to take away with the concept of machine, because it is not capable of building options or routes with the items it has.

You understand that the difference lies in the organization that allows self-regulation to maintain the unity of the microcosm under any circumstances. Understand the above, we can return to the body with more specific questions about psychosomatic disorders and chronic degenerative diseases.

The conception of the body as a result of the historical-social leads to not accept that the body is a machine, which is not divine creation and that, is not predictable in its construction. The response of a body-regulating unit constructed and has its idiosyncrasies put us away from deterministic processes in any fields. To open a reading of the body not just accept that is the result of a process which is able to regulate, it must go to other areas such as language, emotions, spirit, inter-relations constructions produced by the body. We recognize no other space where they develop, grow, germinate, harvest and store individuals. This confronts us with a different problem, the discussion of what it controls, subjects, develops, obstructs, or ill the body or if it can reduce human activity to the realm of language, consciousness, thought, the unconscious or behavior. Discussions on this process can lead us to look at the outward expression of body and divert us from the search path is constructed in order to maintain unity, but sometimes destroys it. Thus, the principles of life and death are in the individual, the point is how to remain in the control unit to produce regulation and, when lost, which maintains the principle of life.

We conceive that the body produces, builds, develops and plays, has a principle of cooperation is the support of the microcosm, where nothing is free from the effects of an event. It is not easy to miss the interconnection between the organs, nor the relationship with the emotional process that is taken into the body. Thus, any pain or emotion that involves the threat of joint cooperation makes you activate a process of accommodation and adjustment to maintain optimal physiological response that preserves the ancient principle of life. The body, then, is life, because cooperation is what sustains life, among other things, can produce an infinite number of processes, is to stay as a unit or embedded in the social, historical, artistic, ethical, religious or scientific. In the rush to produce explanations, science, technology, poetry sublimate more abstract processes. This sophisticated organization of the body allows us to say which is the only body that can be as dangerous or holy for congeners.

Thus, a body makes its condition, its state of life in clear relation to the living spaces and times, the result of the preceding culture in its time may to exceed or to perfect, because it is solicitor of new body processes to be weakening, destroy or strengthen the unit.

If the biology of self-regulation can not be taken to the emotional process as a factor that crosses the cell and organs, then our eyes will be truncated and prevent us from explaining the topic of interest. Weighing the cellular process do we deny the emotional process? To our eyes, emotions are part of the unit that regulates the body, even they have a role in contemporary processes of construction of new signs and symptoms that allopathy considers sequential laws and principles that have been met without rerouting in the last three hundred years. But now you can find all the symptoms of a heart attack without any myocardial injury record, or gastritis or colitis without necessarily mediate lack of food, which is then attributed to stress, "the nerves." The fact that an emotion can deregulate the unit body means the body has routes of expression that allopathy has not been able to identify precisely by conceiving it as a machine. By contrast, the body itself can build, develop, transform a physiological process that is believed established or determined by social conditionality, the surprise is that this way of life can be built via the emotions, symptoms and signs that are indicators a certain disease to be medicated does not improve, indicating four important moments for the body:

1. The message body is built wrongly interpreted; a reading from the causality only justifies the fragmentation and partiality of the solution for the cure.
2. When medicated, the body has to develop another mechanism to process the drugs, that is, there is more work for the body, which implies an alteration of the energy economy of the individual, and may collapse or damage the control unit of the body, then get sick with another disease. This means that gets violent on the principle of cooperation in the sense that this principle will have to exercise its organization to keep it violated, but that in turn becomes another problem. For example, if the answer to solve a problem is vasoconstriction, the effect of this response will affect the heart, you will need to pump more and therefore grow and reduce its ability to pump. Hypertension is the result of the breach

of the principle of cooperation, which in turn will affect the kidney and its parallel responses such as headaches, eye and dizziness, which enable emotional states of distress, anger, depression, sadness, and so until the body dies.

3. A body with this load cannot generate a regulatory production in a society that has the culture of anesthesia (the ordinary is trying to remove the pain without asking what is the subject doing wrong). Not feeling the body becomes a state of denial that expresses a fragmentation of the biological and emotional. The implications of this are living in denial the clinics: people live split and inhibit the pain only hides the symptoms as the deterioration or degradation of the body continues. This illustrates, for example, the case of arthritis: the administration of cortisone inhibits only the pain but not cure, and while the patient does things without pain, the dose increases to inflame the face and body parts. By then the pain will be unbearable, the body will continue protesting to preserve the principle of life, despite the large doses of medications continue to build networks of survival with a body that functions optimally. This is a great lesson of the body for understanding the strength of the network of international cooperation that we living beings have.

4. The importance of the emotional process in the body and their explanations have diversified, are fields of knowledge that do not touch in diagnosis. An atomized conception prevents the acceptance of other explanations for understanding how emotion travels in the body according to a principle indivisible of unity body-emotion. The fundamental problem to accept this idea is that emotions are explained from proteins or neuropeptides that are present when an emotion is expressed. In our logic we can understand that these substances are the expression of a dominant emotion that encourages the production of substances to affect, by its excess, to another organ, and even encourage the expression of other emotion that relates to another body.

It explained that the prospect of the unit body-emotion could make it understandable or predictable about what will happen to the subject. We can illustrate this with the case of diabetes, which is the topic at hand. We start from the idea that diabetes patient's report that they were told from an emotion like fear, anger, anxiety, or a mixture of all them, which allows this a predominance of the emotional in the unit. The intensity which individuals live this process exacerbates the representation of a balance in the body. People carried to an extreme emotional condition without the realization of an action, this mens, you only live the excitement in physical space but is not expressed outside, and that generates work processes in the organs to bring into being or self-regulation, and this is where the body's condition has changed. While that is capable of regulating itself to meet the conditions under which it is, their responses to exacerbate this action is an emotion that should be regulated, to overflow the equilibrium condition of the organ-emotion making possible that some organ work at top speed, thus undermining their inner workings. The body is subjected to an energetic effort to return to the condition that it altered. Should be clear that what happens to the body is not the result of Darwinian evolution or a simple transformation of social conditionality that prevents us from having another look. Culture and history of the emotions play a role in this new construction of physical reality. The body is crossed by the customs, laws, morals and ethics, which establish patterns of behavior, social regulation, conditionality, which implies that the socio-historical process is present in the body in daily life. The institutionalization of the body can become violent when the management of emotions beyond the control of human relationships., That emotion can not be disjointed of the historical-social of the body as an expression of the unit body-emotion, and this means that emotions have a major impact on the functioning of the organs. The relationship of autonomy is not conceivable in our reading of emotional processes.

Stress, anxiety, gastritis, colitis, hypertension, tachycardia, headaches, etc., Confirm that emotions have effects on the body, but it really is not incorporated daily as the expression of a more complex reality, in any case is defined as natural or normal, when what happens is that it has lost control of a unit that is regulated under adverse conditions. The predominance of hours, days, weeks, months and years of that condition, and exceeds alters the organization of the unit, disrupts the harmony is created then a new body in their organizational processes that have to survive or build deceptive subterfuge to network or red flags that the subject can not read.

The body sets new survival networks not to let the unit dies. This energy effort, only enables a new process that makes the individual age, get sick, build autoimmune responses, which the condition becomes chronic, the body has periodic crisis. It then can be identified as a psychosomatic illness or chronic-degenerative.

Emotions are an unavoidable part of physical reality, so we cannot look for explanations outside. We should go to the unit process organ-emotion and understand how emotion is transformed, as it changes to the detriment of the individual.

It's worth noting that in the balance establishing a harmony that maintains the health of the subject, but we must not lose sight of dimension of a culture that can generate fear, terror, melancholy, sadness or anger, which means that social groups will see with normal persistence of certain degenerative processes. For example, it is expected that someone is diabetic because before someone was in his family. This prediction may be fulfilled in self-affirmation with most people because it provides a way of life that regulates compliance (a hereditary factor is culturally important the dietary habits), so it is expected anger or fear as detonators of diabetes. This construction cannot be appreciated without social and family history, so you can accept it for granted[195]. The unit body-emotion is a unit that when losing its balance builds new networks that are not always the most beneficial for individuals.

With these thoughts and considerations on the process of emotion and its implications for organs is that we approach 74 women with diabetes to find the emotional process of a body that allows excess glucose in the blood tissue.

---

[195] When people are angry or afraid, the liver produces glucose in large quantities as a system that enables massive energy (sugar) to defend, run, attack, etc.., Which not being consumed, causes imbalance in the production of insulin in the pancreas.

Some results and analysis of the body-emotion relationship.

The blood glucose values considered normal has changed over time. Some labs even interpret it measures range from 110 to 120 and 130 mg / dl, which justifies the existence of prediabetic, asymptomatic patients and, of course, declared diabetic.

The 74 patients have reported higher rates to 130 and have reached up to 400 or more, which means that they have parallel problems of eyes and kidneys, to which should be added that 33 of them have hypertension. All reported having diabetes from a period from the last 15 days to 22 years. Other related conditions are muscle aches, malaise, fatigue, and emotional states associated with anger, complaints, disagreements, worry, anxiety, fears, depression and sadness with a dominant presence in their daily lives. That means that the unbalanced emotional process helps individuals increase their glucose emotional management. If we consider that there is a dose of glucose when angry, scared or anxious, it means increased rates of blood glucose because the pancreas cannot regulate a wholly excess glucose in the blood tissue.

Another important factor to consider is that patients have a dietary history that is important to discuss the regulation of the production process of the hormone insulin. Their diet consists of food rich in animal fats, sugars, white flour and animal protein. That means the pancreas has been overworked since childhood, which helps to inhibit the production of insulin, which the passage of time causes imbalance of blood glucose. The vast majority of patients have reported diabetes after 38 and 60 years of age.

Eating habits are vital to build the unity of body, emotion and self-regulatory process. All patients reported having or breakfast left overs from the day before, that means that the nutrient quality is poor, more starches, sugars, carbohydrates and fats, which means more work for the pancreas, since it can be metabolized, in the time required, there is excess sugar in the bloodstream. They are further characterized by not eating raw or steamed vegetables, if eaten come to eat boiled or cooked too, which means zero nutrients. Fruits are absent (see Table 1).

Breakfast foods are rich in sugars, starches and fats, which means that the production of insulin is working at an early age. The midday meal is rich in animal fat, starch, sugar and water boiled vegetables, at dinner the food back again with sugar, milk, bread or stew noon. Note that this food will turn to eat at breakfast the next day, which means that the food consumed is almost entirely light brown, then we can conclude that the basic colors of the food is not present in the diet, but rather there is an imbalance in food intake in the absence of products in purple, red, yellow, green, blue, and so on. We can see that there is no concern for what is consumed in the quality and color of food, which is understandable given the conditions of economic situation. However, that cannot be reason to eat foods with more than twelve hours of preparation.

The sense of eating is linked only to fill the stomach. This type of nutrition helps the body to wear that over the years has serious implications, which conjugated sedentary lifestyle contributes directly to store lipids, and sugars that are to harden arteries with all the risks this implies for the heart, brain, oxygenation, and so on.

TABLE 1. Common food habits of 74 patients.

Breakfast	Lunch	Dinner
Stew taco from the day before, celery smoothie, nopal, and grapefruit. Xoconostle fruit, yogurt with apple or oatmeal, cereal. Egg, soy milk, oats, amaranth, raisins, nuts. Fruit, sometimes integral sandwich, coffee with milk, bread, maize drink, eggs, beans. Panela cheese with crackers.	Red meat, cooked vegetables. Not eaten raw. Beans, pasta soup, rice, chicken, fish, tortillas, water. Rice soup. Stew chicken, beef, fruit, (sometimes). Steamed vegetables, raw. Very scarce fish. They eat light products. Soda.	Bread, cereal, fruit, coffee or oatmeal, or half of sweet bread. Sometimes toast. Milk with banana, milk with coffee, cookies, bread, roll, or do not eat dinner.

Table 1 shows the overlap in food habits. The education of the palate is related to food "tasty" or "rich", which means foods high in fat, sugar, starches and animal fats. There are no plant foods and fruits, and it is noteworthy that although people with diabetes do not change their eating habits. To this we must add the emotional process of daily living.

Patients have built an explanation of diabetes from an emotion like anger, fear or courage; 80% of patients report this condition. It should be noted that emotions are in moments where subjects have to defend, attack, and to retaliation, to prevent diabetes is present. Patients identified a formal logical association with body changes. Their diabetes-emotion representation allows us to know a different process with unit body-emotion that affects the possible relationship of liver to pancreas in the second relation of the theory of five elements, kidney and pancreas in third relationship. What will change the relationship between the organs is a strong impression by accident, fear or loss. This matters considering that they all had a violent impression that changed their body's metabolism. For example, the subject and body should feel threatened and run, fight, become angry, which means that the subject uses his body to defend himself, this means, self metabolism enters a complex process of survival. This involves the construction of new answers that are linked to diabetes as they create new networks of destruction that combine with the sugar-rich food history.

We cannot say that only emotions are what transform the pancreas; the relationship between bodies and emotions allows us to give a different meaning to the process of experiencing diabetes.

Having high levels of glucose can be accepted as natural because their ancestors had the same. The explanation and acceptance of the above is related to previous health problems, including constipation, headaches, digestive problems, colitis, gastritis, sore bones, but especially, all reporting an emotional state that includes depression, sadness, anxiety, anger, complaints, insomnia, etc. (see Table 2) state that once declared worsen diabetes. Also presenting recurrent peripheral facial paralysis in the face. The complexity of these relationships, emotion, food and body gives us the key to personal construction in which there is no information about the disease and its implications, especially if the rates of glucose are high permanently.

These symptoms are messages given are expressed as the construction of the cooperation network that tries to solve a problem, but another problem arises and deteriorates the body on a principle of self-regulation that works from the principle of preserving life. Many of these conditions are related to diabetes at different stages. The answer is expressed cooperation in solving a problem, but it appears another and creating a vicious cycle that culminates with the destruction of the cooperation network. Of course, the link with the emotional process takes just with the increase of glucose; the emotional state of the women studied is usually anxiety, sadness and anger, among others (see Figure 2).

TABLE 2

Most common ailments of 74 patients
Loss of sight of one or both eyes, retinopathy, glaucoma, fluid retention in knees, feet. Kidney failure, urinary tract infections. Pain in hip, both sides of back, breast, feet, waist, knees, neck, heels, and sciatic nerve head, cold feet, cramps and numbness in both extremities. Fatigue. Neuropathy, dry mouth, extreme thirst, insomnia, cholesterol levels and high triglycerides, hypertension. Burning in the back, colds, cough and recurrent tonsillitis, loss of appetite, swollen abdomen, gastritis and colitis, diarrhea, hemorrhoids, constipation. Osteoporosis, ringing in the ears, heart attacks, varicose veins, herpes skin infections, vaginal hysterectomy for fibroids, cyst in liver, dizziness, facial paralysis.

High levels of glucose are articulate and disarticulate with the lifestyle. We found that women understand the disease as a lifestyle, it is expected that this happen, which implies a contradiction. The diabetes study shows a close relationship with a process of building new networks of self-regulation to make sense of a life project. The expression of the fragmentation in their bodies leads them not to find links to the disease, implying a gradual destruction of the body of an unconscious way. That leads to a way of seeing and experiencing the process of glucose, resign, become indifferent; they enter a process of self-destruction.

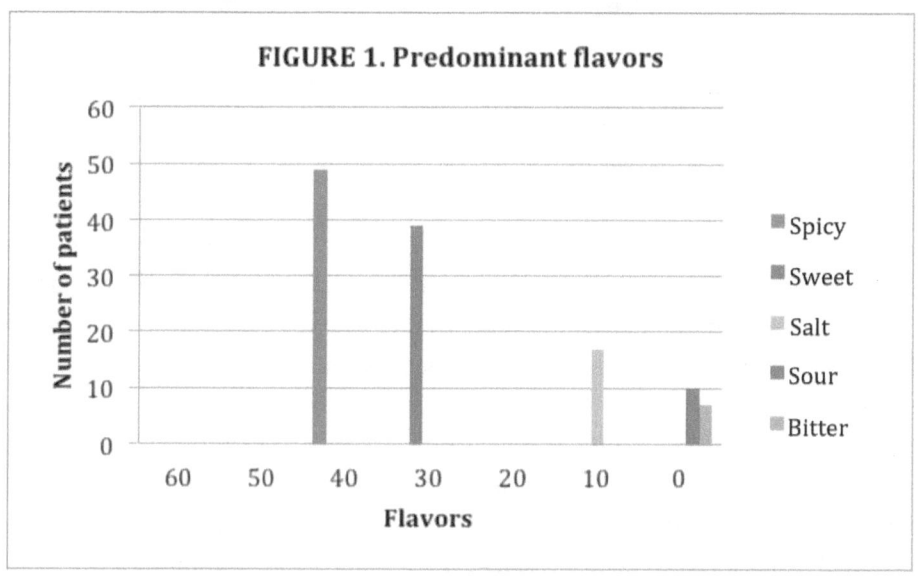

FIGURE 1. Predominant flavors

Anxiety levels are characterized as a rule of life that is expressed in a troubled relationship, whether with your spouse, children or other relatives, so the leads, therefore, to have high levels of glucose. So acupuncture treatments allow you to calm down and can be viewed from another perspective, make contact with the body and define their care. However, the food factor obscures a change in people, which means that the bodywork is not present. The fragmentation is expressed in the rates of glucose.

Figure 1 shows the predominant emotional state through taste for flavor. The predominance of the hot shows that are sad 49 women, 39 of them show anxiety about the sweet, 17 of them show fear to prefer the salty; the sour, linked to the liver, denotes anger in 10 of them; and the lower, with 7 adepts, is the bitter, which is related to the heart. The sum does not match the total number of women reported preferred because they wanted more than one flavor at a time. The taste is a reflection of the emotional state that is the highest expression in the body, allowing you to identify the predominant emotions. All this means that the body is immunosuppressed, has more appetite and desire to eat sugar or foods rich and tasty. This information is important to understand why conditions are articulated with the symptoms that express only self-regulatory work of the body. It should be emphasized conditionality that causes normal or natural to see what is not: their ancestors lived well, for example, and they expect the same. There is, or is not, the ability to work to change the story or memory of the body.

Therefore, it is also natural to be the dominant flavor in personal life, this means, the construction of taste is an identity, a way to live the relationship with food and emotions. We can illustrate this with the spiciness and the correlation with the emotional state of sadness. Its prevalence makes sense with the flavors that follow, as the sweet, and even the figures are reported more numbers because they like two to three flavors, which tells of a vicious circle: to be sad and eat something spicy, be anxious or worried and eat something sweet, to be afraid and try something salty, etc. The joint food-emotion-taste is related to the intersubjective relations with other representations that prevent the possibility of living the body and with the body. This realization of a reading of fragmentation does not allow other possibilities to work with the body and emotions, the determination to say, "Well that is how I am" "and so was my ancestor of first and second generation" contributes to building self-regulation establishing as impossible to change.

In this process the subject has an important role in giving meaning to their lifestyle and move into the world as constructed according to the truth of its representation to the body and their ancestors. It may not be able to conceive other views on life styles or feel the body, even as the space where you can build a new way of feeling, so the chart of emotions most frequently indicates a conflictual social life, predominant emotional state and not only dismantles and unbalances the body, but also determined ways of seeing the world in disarray and with a sense of resignation. Is that what "normal" becomes a pathology and nonsense in the body, this means, become victims of circumstance and that does not open any new possibility of seeing the world and the body as the spaces that we can build and refine. The body falls into disrepair, like the house, neighborhood, municipality, district, state, and country. The most common patterns of disease indicate a process of bodily deterioration cannot be stopped when the process is advanced, this means, the body wears the new outlets to the destruction process of self-regulation. The responses will be containment, but also new routes, so we can find patients with high levels of glucose go walking though logic tells us that they should be hospitalized. To understand this phenomenon of patients living with rates above 500 mg / dl or, conversely, that others who have more insulin to be administered and cannot lower their rates, involves understanding the conjugation with unresolved emotions, containment because invariably have consequences in peripheral areas related to a system or organ. This reality can be understood as a process of building that fits perfectly with the Taoist theory of five elements, which confirms that fragmentation is an obstacle to understanding the new routes, which are building a body.

Figure 2 gives an account of a lifestyle in which sadness is predominant. 59 women sad and 32 angry is an important fact to understand the emotions that allow elevation of blood glucose are related mainly to sadness. The fact that most of them recorded sensitivity, fear, worry, nervousness, crying, depression, irritability, anxiety, melancholy and despair, and only 6 are felt happy and peaceful, reveals an alarming emotional situation.

FIGURE 2. Frequently emotions

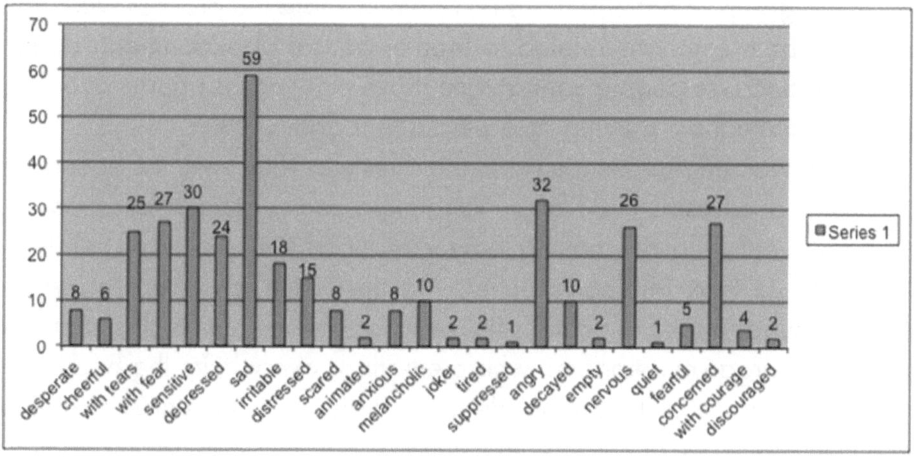

Emotion

These emotions may well illustrate a process for balancing interpersonal and intersubjective relationships, so we can understand that their glucose levels are not regulated as easily with a drug that cannot regulate ingestion of foods rich in sugars, starches and fats. This conjugation with the emotional state allows us to understand that a body that attempts to regulate cannot build other options in the space it inhabits, this is, and the process of coexistence is not pleasant. It is significant that only two of all women who studied are jokers and other two feel cheerful, six can say that happiness itself exist in their lives, but outside them the emotional state of the population does not contribute to their health or the self-regulation process to preserve life. So it is not possible to explain the increase in glucose is due to only one type of food, because it also ties in with an emotional state. The figures are higher because they have more than four emotions in their body, moved from sadness to tears, with weeping to depression, so that we can understand an increase of problems.

The body harmony does not exist in these female bodies studied. Would have to find other ways of understanding relationships and linkages of these emotions that do not accommodate to the joy and the joke, the mood for a better life. The circle of destruction of the closed space of the body while increasing emotional state does not change, indicating that it should work with emotional states, learn to smile, be happy; change, then, the attitude toward life.

In the group of patients there is a path to share the process with fear, anger or both emotions related to fear of abuse, abandonment, death, violence. Because people cannot face those emotions properly, overflow and can make them a living or a permanent grievance.

These women represent a particular lifestyle. They lived the economic and political crises of 1981 and 1994, experiences of violence, assaults, robberies, etc., the purchasing power reduced the quality of nutrition, which led to look for food "richer" "tastiest" rich in sugar, fat, white flour. So sugar intake at home is linked to anxiety as part of this complex process of organ-emotion unit where emotions overwhelm the body and affect the pancreas. Changes made to the subject may be further specified within the network of organs, that means that the body of the woman faces social stigma, segregation, to the ways of submission in the family relationship with husband and children, all this allows us to envision how a body decays in a relationship that is not healthy. Living with the exaltation of emotion such as anger, anxiety, fear, sadness, worry or melancholy favors only the rates of glucose does not go down so easy, and in these circumstances you have diabetes or a woman can be an indicator of lifestyle altering the process of insulin production. The rates of women with diabetes are part of a complex process characterized by the subject and emotions. Nutrition can be an option in combination with the physical exercise to make sense of a person who can be free of emotional overflow that makes you a candidate for diabetes. Of course, social and personal status play an important role in the ways of living the body, or if you will, the unit bodies-emotions in a clear linkage to the process of lifestyle and intersubjective representations. Women living alone, widowed, separated and married seem to have the same emotional conditions those sisters in the condition (see Figure 3).

The combination of eating habits, emotional abandonment, claims, dissatisfaction with life, bereavement, problems with children or spouses, is a risk factor that brings into play the indices of glucose.

FIGURE 3. Marital status of patients

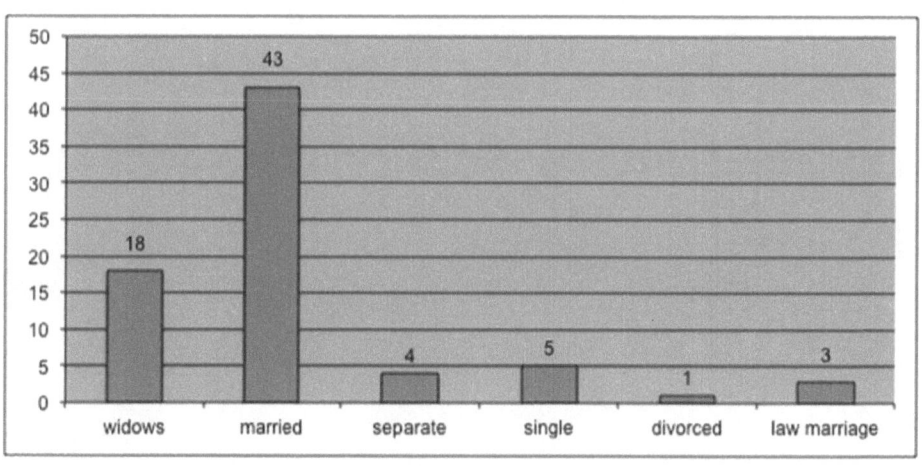

Number of patients and marital status.

The question is: what is the benefit of being in a marital or another status. In fact, the great majority of married women have problems with glucose control. That means that family problems are built into the living spaces and joy does not come to them. Sadness is a factor that crosses either. We understand that emotions are not derived from marital status, which means that the process of construction articulates with ways to live a lifestyle. There will be understood that sadness and anger crossed all these women mostly. We should suggest that what happens in the space of the house is a deterioration of the ways of living, this means, intersubjective processes have deteriorated as well as inside the body. Taoists say that what exists within the body exists abroad, married women and widows are the expression of a deteriorating family life and is expressed in the rate of glucose and the diseases specific to a process that dignifies home life of these women.

IV

Conclusions

I

Diabetes is not difficult to explain, not a curse or a punishment from God, much less an inheritance or biological determination of the ancestors. We leave the body as the only place where you build the self-regulatory process of life. Our objective was met by showing the relations of bodies, emotions drive a body building process on diabetes in the body of women who were treated at the Center for Psychological Studies and Attention A.C. from the epistemology of acupuncture, which conceptualized the body as a microcosm that is articulated in a unit related to the elements, wood, fire, metal, earth and water, which in turn are linked to the seasons, emotions, growth, development, harvesting, germination and what is stored in the body, and of course, social and individual processes that can live in the family. What underpins this is the theory of five elements, from which analyzes the relationship that integrate nutrition, emotions, and the predominant flavors in conjugation with intersubjective relationships and their link to chronic diseases and psychosomatic more frequent. The concept of unity between body and emotion is an approach that maintains that there is one without the other, and are an indissoluble unity. This logic I hold that the human body should not be fragmented for its study in any conditions. Thus the process of the indices of glucose with and without control this process relates to the construction body and intersubjective. We see that the self-regulatory body is not in the line of building new relationships in search of perfection of a body process to be a place where patients can feel or build new options to live their emotional processes.

Breaking the emotional vicious cycle that builds each demand a personal work process that is complex relative to the clarity of seeing the world and social relations that are complicated and codependent for women. To this we add the educational factor in the sense of learning to see the world and explain personal and social processes, disjointed or related to its body process, because everything indicates that the distance from the social and the body is abysmal. Reduce the process of a diabetic to an emotion that triggers the disease to build a truth that cannot be broken before the tangential evidence of the fact a prediction is confirmed by itself called diabetes. The flavors are a clear example of this process that allows food joint with a culture of food at home. We can say that the traditions and customs of the family are not as healthy as you try to believe in wanting to exalt with a value; eat barbecue every Sunday is not a tradition of pride or denoting economic capacity, it is rather the result of body building inevitably show its effects years later. Consume sugar and starch foods that are tasty and rich, shows that only the perversion or prostitution of the palate leads to the addition of the spicy, the sweet, of salty. No matter the social and marital status of patients, the game of emotions is present. The ways of constructing meanings and representations of the body only allows the process of self-regulation be a different construction to harmonious life principle, this means, survival pathways are activated, and are giving way to a new body, which is related with politics and economics of Mexican society. Eat *light* foods also stands to take the disease, as a pretext to acquire distinction or social class thus becomes a space to build roads that do not change in terms of symptoms. But one thing is certain, we find that the flavors and emotions are linked to the body building process that make these women in their rooms in the house with a heritage that is shared food. Well may we say that education of palate can be explained with the latest ways of living the body. Diabetes can show that the ways of living at home are an important point of the disease. We can get lost and the problem of diabetes is just the rate control blood glucose, but now reported indicates that the problem is not so simple, it shows the background rather than a look of the disease indicates that it is need to work on other routes of understanding of diabetes process that expands in all Mexican households. Women are the architect of Mexican society, and if so are they, the formation of the children will have a seal. The process of prevention should go to work with the body and its implications, and to think of it as a space that can be regulated, built and is not determined or sentenced to receive the rates of glucose that will inherit the ancestors. The research area is open on this route to think about prevention.

II

The looks presented here are a point, an interpretation that highlights the need to learn from what we face in our country in the diet, especially in a lifestyle that becomes the norm for everyone. Being a diabetic today can be understood as a natural consequence of an inheritance, a condition expected or become a test to measure how strong we are. Today the diagnostics have become commonplace to make sentences and exclusion for a lifestyle, the search for alternatives are not foreseen, excludes any alternative. The impact of a voice like that of doctors is damning: the doctor said, specialist, responsible for health. This authority does not cease its proposed options to "eat well or take care of your diet." Humans make them a life style brand.

What is done with the body of diabetics is not free of the interests of any kind, from marketing to the pharmaceutical industry showing the claws to the inhibition of the production of insulin in the pancreas, the pain of family members, dynamics that are built when the patient enters the phase of dialysis, neuropathies or search for a kidney transplant. The complexity of working with diabetics not only tests the filial love or work of someone in particular, but also establishing routines and styles that change the ways of being of people. The complexity of human relationships is not the only thing at stake: the interests and love are combined with the guilt and remorse. Diabetes also questions the partnership; body dropouts are not only part of a culture but also a logic in which a body is built with the idea of body inheritance in social terms. The sentence becomes hereditary determinism is a theory to be charged for generations. In this logic there is no exit, you cannot build another option other than premature death.

We believe it is necessary to look into new forms and styles of living in the same space, to become builders of our body and we must take some steps. The most important is to have accurate information on how to avoid it even if we have in our history diabetic mothers or fathers. For us it is increasingly clear that the construction process of diabetes is tied to the process of a culture that we introduced a stylish look and feel, with flavors and tastes that are said are part of our culture. We must open our eyes to these traditions and customs that contribute to a deterioration of the organs, changing food habits, culture and style that create an imbalance in the body's metabolism. The feeding is crossed by the beginning of what is not rich yet nutritious and taste of palate is prostituted. This complex web of the disease makes the body impossible to prevent. Will our age of information has become the era of ignorance. For the same diabetic or are concerned about what to do or not with respect to treatment. Placing the right to health workers is to avoid a family problem that is present and take away from a body and a social history that we can explain what happens in the present.

Knowing the culture and history of the society today allows us to give meaning to what we as a body, a body full of conditions that destroy it, just see the statistics of epidemiology, treatments and outcomes, indicating that despite all, death due to secondary effects of diabetes is about as common in our time. Given this reality, remedial medicine may not be the only option. The power of information just give us the possibility of conceiving the body as a space from which to build and cultivate life in balance with emotions and social processes, factors in the medical discourse not ever articulate in explaining diabetes.

Maybe we should make a consideration: the body cannot be seen as a machine, we should not feed the deeply rooted belief that the body needs a replacement. No, you must change our conception of the other body on which it becomes clear that the body is the epic, builds, the farming life to make sense of the actions of a generation. It should be clear that the body is capable of regulating and even reverse processes.

The research aims to build routes we travel from one emotion to observe the place where he is staying to unlock or disinhibit what is stuck. The process is complex, but not so if we learn to see the body as a machine, but as a space that is cultivated early in life and are constructed ways of being and of dying.

Human beings are not perfect machines. We understand that the bodies of today are sedentary and eat too much sugar and carbohydrates, animal protein and starches, this type of food affects the pancreas but also the heart, kidneys and lungs and other organs. Food is the main imbalance that prevents people from having a better quality of life.

So what is our alternative? We can start from a change in feeding and regulation of a constant and ongoing exercise to allow for oxygenation and removal of toxins and fats. However, know people who have that system and still develop diabetes or cancer. In reality the emotional process is the focus of this new body that is part of a new physical reality. That's what we mean by a "new body", a new reality, no doubt this does not coincide with the established, but we must learn to see the world through different eyes.

Many answers are in the body; only epistemology that dominates in the middle needs to be updated. These are new times, figures and statistics tell us that something is going wrong, surely cutting measures are not epistemological, and demand a design that truly respects life in all its forms, so that can be reflected in the existence of future generations. Otherwise, all that is supposedly for health will only simulation and our descendants will die in the same or worse than now. In any case, we have the option to take responsibility for our lives and take care from now on.

20

Conclusion

New times live inside the body. The relationship body-emotion has become a new space within the body, a unity that goes beyond the archaic idea of mind-body and even the proposal that emotions are a fruit or social representation, the process has become complex in the eyes of the functionalist or positivist and even looking image representations. The complexity of the body is a reality that we have in our pores, digestion, insomnia, dreams, skin, uterus, penis, etc., Have taken over the various body spaces and have imposed a rhythm, a movement styles of living in megacities. The boundaries of stress, anxiety and competition, have incarnated, have become part of everyday life of individuals, the column deforms them, makes them colitis, gastritis, migraine, dermatitis and if were not enough, away from the idea of happiness. Perhaps the above is not new to say, but if the fact that all human beings aspire to that sector of the emotional body processes dominate. Mexicans are not only those suffering from this process we have assigned the name of the body, which has its history, memory and away from the evolution of the idea of divine creation, belonging to a class. The body is a living space where all emotions ancestral human beings have feelings and are building a society that seeks identity as a group, we, in these 30 years of work, we have found that individuals make use of their choice according to their social desires incarnated in the body, this means, artificiality expressed a desire that never met, which leads to suffering for want to have, keep, care for, cultivate and nurture the idea of having allowed losing the middle path. That's where we find, that an emotion has become a sort of fugitive from the organ to which it belongs and is expressed in a false autonomy.

The expressiveness of the body is a horizon that has nowhere to go, only the body by responses to survive, to endure the dangers of a society to its health. The body defends itself under the pressure of competition, envy, but is complicated. For more answers to develop, its scope for action is the carnal. Its buildings are complicated and complex to repel the aggressive out there, which means that the inside is under threat as within so without. Only the artificiality of our lives in society does not allow seeing this physical reality that is sophisticated in the development of chronic and cancer.

We have in our eyes the expression of a decadent society, which eats away the body with the loss of respect for life. It does not matter the nature of our body, it has lost the sacredness of existence, both in the individual ceases to be sacred internal and external, any dialogue with the elements around us and within our interior.

Our body space has many lessons well worth it to practice and learn, know your body and not only cognition, are the new challenges for pedagogy of the body, where health is a bodily practice aims to go prevention. That means we aim to self-sustaining body, although we understand that the social and political organization of contemporary societies, are designed to consume bodies and nature, the monster eats bodies, which is the consumer society does not give us options rather than take responsibility for our body process, be self-sustaining should not be a political *cliche*, is the act of working with the body, move your emotional memory cell is to change habits, to transform the movement inside and put into perspective the construction for the future, self-sustaining life beyond the body itself, example, to recover the planet as our house, the balance was broken by desire, dominate, control, possession, fame, money, power, etc.., that is destroying the body of our time.

The reason why we did all the research of the past and reflection of today to do something with the body of present, but yes, the responsibility we have to work it into the body so it is not an ideological discourse that seeks to justify laziness and inactivity that are killing the body of today.

21

Bibliography

Aceves, Jorge E. (Coord), *Historia oral, ensayos y aportes de investigación.* México. CIESAS.2000.

Adams, William, *Las raíces filosóficas de la antropología.* Madrid. Trotta, 2003.

Arendt, Hannah, *La vida del espíritu,* Barcelona, Paídos, 2002.

Barfield, Thomas (ed.), *Diccionario de antropología,* México, Siglo XXI.

Battan, Ariela. "The paradox of the distinction. The problem of soul and body in Descartes and Merleau Ponty." Journal of philosophical history. National Autonomous University of Mexico. Institute of philosophical research. N3, p.13-15

Camarena, Mario, *Jornaleros, tejedores y obreros. Historia social de los trabajadores textiles de San Ángel.* 1850-1930. México, Plaza y Valdez, 2001.

Campillo Álvarez, José Enrique *El mono obeso. La evolución humana y las enfermedades de la opulencia: diabetes, hipertensión, arterosclerosis.* Barcelona, Drakontos, 2004. 235 p.

CAPRA, Fritjof, *El punto crucial ciencia, sociedad y cultura naciente.* Argentina Estaciones. 1992.

Capra, Fritjof, *La trama de la vida.,* Barcelona, Anagrama, 2000.

CERTEAU, Michel de. "Historia de los cuerpos. Entrevista con Michell de Certeau." *Historia y grafía.* Universidad Iberocamericana.

Chávez, Ezequiel A. *Rasgos distintivos del carácter mexicano.*

Cuellar Romero, Ricardo, *De obrero "músculo" a obrero "intelectual". Modernización de la industria textil del algodón en México alrededor de los cincuenta*, México, CEAPAC ediciones. 2004.

De Hipónoma San Agustín *Confesiones*. México, Alianza, 1997.

Dejours, Chirstophe, *Trabajo y desgaste mental. Una contribución al a psicopatología del trabajo*, Buenos Aires, editorial Humanitas. 1990.

Devos, Geroge, *Antropología psicológica*. Barcelona Anagrama, 1981.

DURAN AMAIZCA, Norma Delia. *Cuerpo, intuición y razón*. México, CEAPAC, 2004.142.p.

Durán, Norma, "La función el cuerpo en la constitución de la subjetividad cristiana", *Historia y Grafía*, Universidad Iberoamericana,n9,p.19-58,1997.

GILBERT, Paul, "Historia del cuerpo; expresión y libertad". *Revista de Filosofía*. UIA. V.333, n.97, ene-abr.2000.

GODINA HERRERA, Célida. "La teoría de género en la perspectiva fenomenológica del cuerpo vivido". *La Lámpara de Diógenes*. Benemérito de Universidad de Puebla. Año2, v II, p. 32 ene-jun 2001.

LOCKE, John *Compendio de ensayo sobre el entendimiento humano*. Barcelona Tecnos, Clásicos del pensamiento. 138.1999, 61 p.

López Ramo Sergio, (coord.) *Lo corporal y lo psicosomático, Aproximaciones y reflexiones*, 7 tomos, México, CEAPAC, Ediciones.

LÓPEZ, Ramos Sergio, *Historia del aire y otros olores en la ciudad de México*. 1840-1900 Méxco, CEAPAC-Miguel Ángel Porrúa, 2002.

_____ *Fuentes hemerográficas para una historia del cuerpo humano en México*. (184-1899) México CEAPAC 2005.

_____ *Prensa, cuerpo y salud en el siglo XIX mexicano*. México Porrúa México, 2001.

_____ *Zen y cuerpo humano*. México, Verdehalago, 2000.

_____ *Fuentes hemerográficas para una historia del cuerpo humano en México*, México, CEAPAC Ediciones, 2005.

Nueva conciencia, plenitud personal y equilibrio planetario para el siglo XXI, Integral, España, s/a.

PIEPER, Josef, *El concepto de pecado*. Barcelona Herder, 1986. 119p.

RÁBADE ROMEO, S. *El conocer humano.1.* Barcelona, Trota, 2003. 614.p

REID, Daniel *El Tao del salud, el sexo y la larga vida.* Barcelona, Urano 1989.

Ribes Iñesta, E. "La formación de profesionales e investigadores en psicología con base a objetivos definidos conductualmente," "El diseño curricular en la enseñanza superior desde una perspectiva conductual: historia de un caso", "Innovación educativa en enseñanza superior, reflexiones sobre una experiencia trunca", en Javier Urbina Soria (comp.), *El psicólogo, formación, ejercicio profesional prospectiva,* México, Universidad Nacional Autónoma de México. 1989.

ROAMNELL, Patrick, Locke, "Un médico ecléctico." *Humanitas* Universidad Autónoma de Nuevo León. n. 10, 1969, anuario, p. 25-273.

SAGOLS, Lizbeth, "Nietzsche y el lenguaje del cuerpo". *Theoría.* Revista del colegio de Filosofía, UNAM, n 2. nov. 1995.

SAID M. W. *Cultura e imperialismo,* Barcelona, Anagrama. 1993.

SHAYEGAN, DARYUSH, *La mirada mutilada. Esquizofrenia cultural; países Tradicionales frente a la modernidad,* Barcelona. Península. 1990.

SHIGEHISA, Kuriyana. *La expresión del cuerpo y la divergencia de la medicina griega y china.* Siruela, 2005.

SHIPPER, K. *El cuerpo taoísta.* España. Paídos, 2003.

Van Ersel, Patrice y Catherine Maillard, *Me pesan mis ancestros; la psicogenealogía hoy.* México. CEAPAC, Ediciones, 2004.

ORAL INTERVIEWS

2004	Interview "Una promesa de trabajo", by Sergio López Ramos, October.
2004	Interview "Todo debe ser objetivo", by Sergio López Ramos, October.
2004	Interview "El mundo de plastilina", by Sergio López Ramos, October.
2004	Interview "De la necesidad a la creatividad", by Sergio López Ramos, October.

www.ingramcontent.com/pod-product-compliance
Lightning Source LLC
Chambersburg PA
CBHW031832170526
45157CB00001B/280